Pedro Nunes (1502–1578)

American University Studies

Series IX
History

Vol. 182

PETER LANG
New York • Washington, D.C./Baltimore
Bern • Frankfurt am Main • Berlin • Vienna • Paris

Pedro Nunes (1502–1578)

His Lost Algebra
and Other Discoveries

Edited and Translated by
John R.C. Martyn

PETER LANG
New York • Washington, D.C./Baltimore
Bern • Frankfurt am Main • Berlin • Vienna • Paris

Library of Congress Cataloging-in-Publication Data

Pedro Nunes (1502–1578): his lost algebra and other discoveries /
John R.C. Martyn, editor and translator.
p. cm. — (American university studies. Series IX, History; v. 182)
Includes bibliographical references and index.
1. Nunes, Pedro, 1502–1578. 2. Algebra—History—16th century.
3. Mathematics—History—16th century. 4. Scientists—Portugal—Biography.
5. Mathematicians—Portugal—Biography. 6. Humanists—Portugal—
Biography. I. Martyn, John R.C. II. Series: American university studies.
Series IX, History; vol. 182.
Q143.N86P43 512—dc20 95-40410
ISBN 0-8204-3060-9
ISSN 0740-0462

Die Deutsche Bibliothek-CIP-Einheitsaufnahme

Pedro Nunes (1502–1578): his lost algebra and other discoveries /
John R.C. Martyn, ed. and transl. –New York; Washington, D.C./
Baltimore; Bern; Frankfurt am Main; Berlin; Vienna; Paris: Lang.
(American university studies. Ser. 9, History; Vol. 182)
ISBN 0-8204-3060-9
NE: Martyn, John R.C. [Hrsg.]; American university studies / 09

Publication of this work was assisted by a special
publications grant from the University of Melbourne.

The paper in this book meets the guidelines for permanence and durability
of the Committee on Production Guidelines for Book Longevity
of the Council of Library Resources.

Printed in the United States of America.

To my dear wife, Alexia, for enduring yet more of my *descobrimentos* in Portugal and for her support and care back in *Nossa Quinta* while this book took shape.

CONTENTS

PREFACE

This book contains important new material for scholars working in several different disciplines, material that reflects the wide interests of a brilliant sixteenth century humanist, Pedro Nunes. It is important, first of all, for Renaissance history and the rise of Humanism, especially in Portugal and Évora, which was the capital of the world at that time; secondly, for the history of mathematics, and in particular of algebra; thirdly, and related to that, for the development of nautical instruments, of marine science, and of cosmology and cartography, which were given practical application by leading mariners of the day; fourthly, for developments in Renaissance poetry in Latin and Ancient Greek, based on the best of Classical models, and finally, for the history of religion, illustrated by a private commentary on the scriptures by a pious Nunes.

Today Pedro Nunes may be little known outside Portugal, except to scholars working on the history of mathematics. However, he is a humanist who deserves to be well known, a brilliant scholar and teacher who successfully combined the sciences and the arts, and who in his wide-ranging interests and expertise was a rival to the great Italian humanists. After studying the Classics in Spain, Nunes completed a medical degree in Portugal and became physician to the royal family. At the same time he was lecturing at the University of Lisbon on rhetoric, logic and metaphysics. He then tutored the royal Princes and some young aristocrats in mathematics, astrology and geometry, while translating Greek and Arabic originals for a series of pioneer works on natural phenomena, cosmology and nautical science. Well paid as the King's cosmographer, he was appointed by him to a chair of mathematics at the revitalised University of Coimbra. There he established himself as one of Europe's leading teachers and pioneers in mathematics, especially algebra. He was also an expert in cartography, and until his final years he regularly visited Lisbon to advise the fleet's pilots and captains as to how they should use the nautical instruments, route-maps and charts that he had himself created.

His family life is also of great interest, with his Jewish blood on one side, and his very strict upbringing in the Roman Catholic faith on the other, which he shared with his Spanish wife, and which gave him so much comfort in his final years. His close and lifelong friendship with Prince Henry, the first active Inquisitor General in Portugal, seems to have protected Nunes from imprisonment or worse, although his two grandsons were not so lucky, both spending lengthy periods in the Inquisition's prisons. The dramatic event connected with his passionate daughter, Guiomar, who cut up her lover's face with a knife in Court, after he had jilted her, is the theme of a newly discovered Latin epic, printed and translated in Appendix B.

Finally, recent discoveries documented in this book have shown Nunes to have been an active member of a group of intellectuals, teaching and writing under the auspices of King John III. This new material relates Nunes to these eminent contemporaries, in some cases for the very first time. This evidence appears both in the miscellany, which contains a rich variety of works by Nunes, which I recently discovered in Évora, and in poems and letters by his fellow humanists, which appear here for the first time in English. These various works combine to give us a most interesting picture of the cultural life of Évora, at a time when King John and his brother-in-law, Charles V, controlled almost all the known world. In an old part of Évora there still survives a Renaissance fountain, carved out of marble, in the middle of which there is a globe showing the exciting new worlds just discovered by Portuguese mariners, and rapidly opened up to the traders, scientists, artists, poets and historians of Europe, especially to those of Portugal. Nunes, who played a major part in these discoveries, might have often sat around that fountain himself, discussing new frontiers with his fellow humanists and his young nautical friends.

For their assistance in the preparation of this material, I am very grateful to Emeritus Prof. R. E. Thompson of La Trobe University, for his help in translating the mathematical and religious texts, to Prof. Ron Keightley of Monash University for his help in sorting out problems in the interpretation of the religious notes, to Dr Jim Cross of Melbourne University for his help in working out the text and equations for Nunes' *álgebra* in English and to Adrian Kelly for proof-reading the text. I also thank the Australian Research Council for funding secretarial and research assistance, and the Publications Committee of Melbourne University for financial help in its publication. For their help both in supplying and in photographing the manuscript for me, I am very grateful to the staff in the Municipal Library of Évora. Any errors which remain, however, are my own responsibility.

Chapter   One

Introduction

The main part of this book (chapters 5-7) is devoted to the long-lost work on algebra, which was written down, as I shall show in chapter 1, around 1533, by Portugal's greatest mathematician, Pedro Nunes (1502-1578). It was used by him to teach algebra to the royal Princes and young noblemen in Évora.   At that time the city was the home of John III, the humanistic King of Portugal, who controlled Europe's first great overseas empire, and who attracted some of Europe's leading teachers and scholars to his Court, including Nunes, Nicolas Clenardo, Jerónimo Cardoso, André de Resende, Jorge Coelho, António Pinheiro and Francisco de Holanda. The close links between these humanists are brought out in this book for the first time, adding to our knowledge of the culture and educational ideals of the royal Court in Évora.

Central Évora remains a perfectly preserved museum even today, with its royal palaces and Renaissance mansions, its splendid churches and marble fountains, and its Roman temple, archways, and aqueduct.[1] The old city, surrounded by massive walls, is the capital of the fruitful Alentejo district, and lies half-way between Lisbon and the Spanish frontier, fortunately well clear of the normal tourist route. The city is popular, however, with Portuguese visitors who point out that Évora is the only town in Portugal described by a noun (*museo*) rather than an adjective. Some modern houses and hotels have been built outside the walls, but none inside.   In the old Moorish district, one of the narrow streets bears the name of 'Mestre Resende', where that brilliant humanist and close friend of Pedro Nunes, as the new material reveals, had his home and leafy garden, filled with his archaeological discoveries, and where he, like Nunes, was a most successful teacher of both royal and aristocratic pupils.

This mathematical text is by far the most important part of a surprising discovery which I made in October 1990, in the research collection of the Public Library of Évora. Three other chapters will cover the other writings by Nunes discovered in the same manuscript. First, in chapter 4, I shall reproduce the Church Calendar which prefixed his *álgebra,* where modern English equivalents appear beside the Latin terms used by Nunes. This will help in the dating of the *álgebra.* In chapters 3 and 8, I shall give new evidence for the life of Nunes, arising both from his *álgebra* and from a group of sixty Greek and Latin poems that followed the mathematical work. The poems themselves will not be printed in full, as they have recently appeared in print, together with my brief introduction, translations and footnotes.[2] However, some poems and verses will be included in this book, where they provide extra information on his life and writings. Finally, there are five pages of comments in Spanish by Nunes on episodes in the New Testament, transcribed and translated in chapter 9, and discussed in the light of his religious poems. These provide new evidence for his later years, and for his concerns about death and his increasing devotion to the Catholic faith.

The rest of this book consists of a Bibliography, and an Index of the proper names that appear in the Preface and nine chapters. Two tables will be shown in Appendix A, to provide a quick comparison between the lay-out and contents of the Portuguese original and the final work on algebra published 35 years later in Spanish. Appendix B contains a seemingly unknown Latin poem on the theme of Nunes' jilted daughter, Guiomar junior, the vengeance-seeking, knife-wielding *Dama da Cutilada..*[3] An unknown Latin poem by Resende on the death of Nunes' patron, Prince Duarte, appears in Appendix C. In Appendix D there are little-known letters to Pedro Nunes and to António Pinheiro from Jerónimo Cardoso, plus two Latin poems by

Pinheiro, together with my English translations of all four, and in Appendix E there is an epigram by Jorge Coelho in honour of Nunes' *De Sphera*, also translated into English for the first time.

In this book I shall not try to write a full biography of Pedro Nunes, although a brief one will be provided for readers who know little or nothing about him.[4]   Rather, I shall present new evidence for his life and writings, for the use of any future historians of Mathematics or biographers of Nunes. In modern Portugal the most attractive 100 escudos coin bears a portrait of this brilliant mathematician, holding in his hands a model of the world which was being discovered and mapped by Portuguese mariners in the sixteenth century. This was largely due to the practical value of his geometry and algebra, in devising and testing novel nautical instruments and in preparing nautical maps.[5] He is seen as the creator of modern cartography, especially through his works *On the Sphere* and *On the art and technique of navigation*.[6] Besides trying out his new theories with the help of Portuguese mariners and ships' captains, for most of his life he served as adviser to the professional map-makers in Lisbon. Ironically, it seems that Nunes never embarked on a long sea voyage himself, although he was friendly with local and North European navigators like John Dee, and his nautical works were read by Mercator.[7]

By contrast, both of his sons, Pedro Areas and Apolónio Nunes, served overseas and both of them died in India. One of his most brilliant pupils, who became his most helpful research assistant in nautical science, was the explorer and future viceroy of India, João de Castro. Castro's imaginative descriptions of meteorological phenomena and precise observations on coastal navigation proved to be a landmark in the history of nautical science, unsurpassed by any contemporary or later Portuguese *roteiros* (pilots' accounts of their voyages) His original *roteiro* of the Red Sea, dedicated to his pupil and patron, Prince Louis, was purchased for the sum of £60, an extraordinary price in those days, by none other than Sir Walter Raleigh.[8]

As well as his pioneer work on magnetic variation and the divergence of the meridian, Nunes was responsible for four nautical inventions; first, a shadow-instrument used to tell mariners the true North, secondly, the loxodrome spiral or thumb-line, to give them a constant bearing, thirdly, another machine to read cosine values, which was finely tuned by his fourth invention, his 'Nonius', a small mobile scale used to sub-divide scales, as on a theodolite, which still bears his name in the major encyclopedias. His 'Nonius' was in fact a Vernier scale, 150 years ahead of Vernier. Thanks to these many inventions, mainly due to his early understanding of algebra, Nunes turned nautical science upside down and gave his country's mariners the necessary equipment for them to cross such vast oceans and discover and map so many distant countries, long before any other European power could do so.[9]

For the purpose of advising ships' captains and improving their world maps, as well as teaching mathematics to the young Prince Sebastian, Nunes was regularly summoned to Lisbon, even in his extreme old age. In his lifetime he was very highly honoured and richly rewarded, and since then no other citizen has been valued so highly in Portugal.[10] and there must be very few if any important coins in other countries that portray a humanist, rather than royalty or a politician, or local fauna or flora.[11] This reveals the importance of Nunes today, as the one mathematician in the history of Portugal capable of destroying the theories of a very eminent professor of Mathematics at the University of Paris,[12] able to impress so many contemporary scholars in France, Germany and even in England,[13] and able to create a new mathematical term, the 'nonius'.[14] By his example, Nunes shows the economic importance of mathematics as part of a sound education in the humanities, both then and today. If only Prince Henry had continued to wrestle with geometry, algebra and Greek poetry under the guidance of Nunes, rather than concentrate on religious orthodoxy, their country might have been spared his short-sighted persecution of the 'new Christians'.[15]

The writings of Nunes show his expertise in algebra, geometry, physics, cosmology, geography and Latin, and his unusual fluency in Ancient Greek, with regard to both translation and composition. His prose works are in Portuguese, Spanish and Latin, and he translated works written in Greek, Italian and Arabic. He also qualified as a doctor of medicine, and his medical skills were highly appreciated by the royal Princes, especially Henry. As seen above, his flair for cartography and nautical science proved invaluable to many Portuguese ships' captains, such as João de Castro and Martim Afonso de Sousa.[16] He was also a brilliant teacher, both of the royal Princes and local aristocrats, and of his mathematics students at the University of Coimbra. The latter included Clavius, praised as the 'Euclid of the sixteenth century', Manuel de Figueiredo, who became royal cosmographer in chief, and Nicolas Coelho de Amaral, his successor in the Chair of Mathematics. His ex-pupils remained close friends throughout his lifetime, and in the case of the Princes Louis and Henry, very generous benefactors.

Chapter   Two

The   Manuscript

While   I   was   checking   through   about   sixty   anonymous
Latin   miscellanies   in   the   archives   of   the   Municipal
Library   of   Évora   during the   late   summer   of   1990,[17]   I came
across   one   that   combined   a   work   on   algebra   in   Portuguese
with   60   elegiac   poems   in   Greek   and   Latin,   a   Church
Calendar   in   Portuguese   and   some   religious   commentaries
in   Spanish.   These   made   up   a   single   quire,   one   of   several
bound   together   in   this   very   interesting   miscellany.   To   see
if   their   author   was   perhaps   Portugal's   best   known
mathematician,   Pedro   Nunes,   I   at   once   compared   the
algebraic   section   with   that   jewel   among   his   scientific
publications,   his   *Book   on   Algebra   in   Arithmetic   and
Geometry,*   published   in   1567   in   Antwerp.[18]   The   lay-out,
formulae   and   notes,   written   in   an   attractive   humanistic
hand,   turned   out   to   be   almost   identical   to   the   full   work
published   about   35   years   later   in   Spanish,   as   can   be   seen   at
a   glance   from   their   parallel   tables   of   contents   in   Appendix
A.   This   text   must   have   been   used   by   Nunes   for   his   own
teaching   purposes,   while   he   was   tutoring   the   younger
brothers   of   King   John   III   of   Portugal   in   geometry   and
mathematics   (including   algebra),   from   1532   to   1535   or   so.

The   sixteenth   century   vellum   manuscript,   *Cod.*   cxiii/1-10,
is   an   unpretentious,   leather-bound   miscellany,   containing
214   folios   in   all.   The   first   quire   consists   of   30   folios,   or   60
pages   (four   of   them   blank),   trimmed   to   measure   15   cms   in
width   and   21   cms   in   height.   The   folios   are   numbered   1-4,
and   1-26,   the   first   four   containing   a   Church   Calendar   (in
Latin),   followed   by   the   Portuguese   *álgebra*   (ff. 1^r-12^r)   and
the   Greek   and   Latin   poems   (ff. 13^v-24^v).   Some   poorly
written   religious   notes   in   Spanish   fill   the   empty   pages   on
ff. 12^v and 25^r - 26^v.   The   next   quire   consists   of   ten   folios   of
almost   illegible   Portuguese   prose,   covering   various

religious topics, but on smaller folios (14 cms x 19 cms), and written in at least six different hands. A printed and illuminated *Epistola de Vasco Diaz de Frexenal* follows, on eight folios. Although bound with the works of Nunes, this quire appears to have absolutely nothing to do with Nunes or with the first or last part of the manuscript.

The remainder of the manuscript consists of 168 folios of Latin, including many technical words and quite lengthy Greek quotations, on rhetorical and legal topics. The text which was being used at the start was the corpus of scholia to Hermogenes' work on stasis-theory, first printed at the Aldine press in 1509 as *Rhetores Graeci*, in Vol. 2, pp 1-352. The discussion of rhetoric is well under way, the surviving text beginning on page 25. It can be seen on page 62 in Walz's *Rhetores Graeci* (1833), which includes the Aldine pagination in the margin. The first page starts mid-stream, with Sopater quoting a famous definition of rhetoric by Lollianus (in Greek), who had argued controversially that rhetoric within its species was simply a judicial skill. The enquiry then turns to the great Hermogenes, and later to Marcellinus, Minucianus and Aulus Gellius, with case after case being exemplified with extracts mainly from the political orations of Demosthenes. No *exempla* at all are taken from Roman history. In the general discussion of stasis-theory there are about six short references to translations of Greek terms by Quintilian, and just one brief mention of that idol of Renaissance orators, Cicero.

For any sixteenth century student of rhetoric, this would seem quite extraordinary, and even more unusual is the fact that almost all the quotations and *exempla* come from the seminal Περὶ τῶν στασέων of Hermogenes and the Διαίρεσις Ζητημάτων of Sopater the Rhetor,[19] and that there is a quotation from Demosthenes on almost every page. The page size and watermark are the same as for the first quire, which contains the writings of Pedro Nunes, and its small, neat lettering suggests a contemporary humanist's hand, in all probability that of the mathematician. After comparing the letters and common ligatures in his Latin and Greek, I could find no real difference, except an upper case gamma in the prose. Nunes' special love of Greek would certainly explain his most unusual concentration on Greek rhetoric.

The Greek and Latin poems were probably written to be shown to Nunes' friends and patrons, and as a result, the script in both languages is far more flowery and the text is mostly very attractive, with minimal deletions. Likewise for the *álgebra*, the text of which is mostly beautifully written, and there are only half a dozen corrections. By contrast, the lectures are full of very black deletions and smudges, some of them covering half a page, and several words and phrases are crammed in the margin and between lines. The small, very dark script is strictly functional, apparently intended to be read by the author alone; it was certainly never meant for publication. The subject-matter must have formed part of his lecture programme during his first three years of lecturing at the University of Lisbon.

Nunes' ten divisions of topics are closely based on those of Hermogenes, the oracle for such classifications. The Latin titles follow, with Hermogenes' Greek equivalents beside them:

1. *De Coniectura*   στοχασμός
2. *De Finitione*   ὅρος
3. *De Absoluta*   ἀντίληψις
4. *De Generibus Assumptivae* ἀντιθετικά
5. *De Statu Negotiali*   πραγματική
6. *De Transumptione*   μετάληψις
7. *De Scripto et Voluntate*   ῥητὸν καὶ διάνοια
8. *De Contrariis Legibus*   ἀντινομία
9. *De Syllogismo*   συλλογισμός
10. *De Amphibolia.*   ἀμφιβολία

For his subdivisions of these 'issues', Nunes again expands the brief treatment of each by Hermogenes. The lecture allocation of this document can only be approximate, but with about 12 introductory lectures (on the missing folios) it would add up to about 36 lectures, two a week for a 13 week semester. To judge from the staff, it seems very likely that almost all of the students at the University of Lisbon were studying Civil or Pontifical Law, and the mixture of Rhetoric, Logic, Dialectic and Legal theory in these lectures would have been obligatory for them. After three years or so at the University, Nunes was invited to take over the very honourable post of tutor to the royal Prince (from 1532 to 1534 or 1535), and in 1544, he was appointed to the Chair of Mathematics at the University of Coimbra. Like his friend, André de Resende, Nunes was clearly an excellent lecturer and tutor, combining his pedagogic skills with some brilliant research work, especially in algebra and marine science, a long way from the logic and rhetoric of his first lectures at the ill-fated University of Lisbon.[20]

The first page of the poems has the date 1544 in the top right-hand corner, in the same hand as the text below, and 1563 appears likewise on two folios, 12[v] and 26[r], which contain his religious notes. From the position of Easter in the Church Calendar, the date of the Calendar (and hence of the *álgebra*) turns out to be 1533, a date which well suits its right hand watermark.[21] These three dates fit in perfectly with those established in modern biographies for the three years or so spent by Nunes in Évora, when he was tutoring the King's brothers in mathematics, geometry, and algebra.

Chapter Three

Pedro Nunes - Polymath

Pedro Nunes was born in 1502 at Alcácer do Sal, on the river Sado, about 60 km south of Lisbon.[22] His parents were Jewish, it seems, and he was registered as a 'New Christian'. In his later years, it must have been his very high standing with the Royal Family, especially with his favourite pupils, Prince Louis and Cardinal Henry, which protected him from the Inquisition.[23] His two grandsons, however, were not so lucky, being arrested and imprisoned in Coimbra and Lisbon about 45 years after Pedro's death.[24] During 1521-1522, he studied Greek and Latin literature at the University of Salamanca, where in 1523 he married Guiomar de Arias, the daughter of a Christian Spaniard, Pedro Fernandez de Arias. They had six children, four daughters and two sons. In 1524 he moved to Lisbon, and by 1525 he had obtained a first degree in medicine, and an extensive knowledge of mathematics and astrology. As an expert on astrology, he was appointed Royal Cosmographer on 16 November 1529, a well-paid and prestigious position. On 4 December 1529, he was appointed to the chair of Moral Philosophy at the University of Lisbon, and to this were added the chairs of Logic in 1530, and of Metaphysics in 1531. At the same time he continued his mathematical translations and research, and he managed to complete his medical studies, graduating as licentiate in medicine on 16 February 1532. To his credit, it seems likely that the Princes Henry and Louis were among his first patients.[25]

Early in 1531, Pedro Nunes (aged 29) was entrusted with the tutorship of the King's brothers, Princes Henry (aged 19) and Louis, and other young courtiers. Ever since the turgid but extremely popular *Disciplinae* of Martianus Capella[26] had been adopted by the teachers of the Middle Ages with enthusiasm, the *quadrivium* had become the norm for tertiary education. It consisted of Arithmetic, Geometry, Astronomy and Music. Nunes was spared the Music, but was given theoretical and practical subject-matter based on the other three, related to the Mathematical Sciences, especially the rudiments of Arithmetic and the elements of Euclid's Geometry. He also taught the theory of the Sphere, the theoretics of the Planets, part of the Astrology of Ptolemy, the Physics of Aristotle, aspects of Cosmography and finally, the practical use of navigational instruments, both ancient and modern, the latter invented by Nunes himself.[27]

From late 1531 to 1535 or later, Nunes tutored the Princes accordingly at Évora, together with some *fidalgos*, including João de Castro and Martim Afonso de Sousa. It was while tutoring that he must have used both the Calendar, and his Portuguese notes on Algebra, for planning and conducting his classes in Geometry and Mathematics. At the same time he began to compose elegiac poems in both Greek and Latin, but mainly in Greek, while staying with two brilliant Classical scholars and Latin poets, André de Resende (1498-1573) and Nicolas Clenardo (1494-1542).[28] For Nunes to have preferred Ancient Greek to Latin is most unusual and almost without parallel among the fifteenth or sixteenth century humanists, except for those who were born in Greece, or the few who had studied in Greece or Constantinople, like Francesco Filelfo,[29] or, like Pedro Nunes himself, had translated many of the more difficult scientific and philosophical works by Aristotle, Euclid and Ptolemy, and treatises by the Greek medical writers.

Lisbon's Chair of Mathematics was left vacant after the retirement in 1533 of Francisco de Melo, Rector of the University from 1529 to 1533. In 1537, despite strong opposition from most of its staff, the University of Lisbon was moved to Coimbra in northern Portugal, and on 16 October 1544, Nunes was appointed to the chair of Mathematics as head of an autonomous department, although joined with medicine for official occasions. There he taught and continued his research until his retirement on 4 February 1562. Since 22 December 1547 he had held the office of Chief Royal Cosmographer, an office which he was to retain for the rest of his life. This provided him and his family with a very generous income during and even after his lifetime.[30] While living in Coimbra after his retirement from the Chair, Nunes seems to have returned to his poetry, and used it to bewail the death in 1555 of his royal pupil, friend and very generous benefactor, Prince Louis. In all, another 22 poems in Latin and 9 in Greek were added to his private collection of 29 compositions. At this stage he also began using parchment with a different watermark for his note-book.[31]

It is worth noting as well that his friend and one-time fellow royal tutor, André de Resende, had by then left Évora to join him in Coimbra, as a Professor of Theology either at the newly revitalized University, or at the richly endowed Royal College, where Resende had the honour of delivering the prestigious public oration on 28 June 1551,[32] to mark the anniversary of the foundation of the College by King John III.[33] Resende was probably still there when Prince Louis died, on 27 November 1555. Even if he had left by then, it seems very likely that these two humanists kept in regular contact, both of them being closely linked with Portugal's Royal Court, including its last King, Sebastian, the grandson of King John III. Resende wrote several Latin poems eulogizing the young King, and Nunes acted as his Mathematics teacher. On 11 September 1572, Pedro was summoned by the young King Sebastian to his Royal Court in Lisbon, and kept there for three years, supervising the reform of weights and measures, and giving expert advice to many Portuguese navigators and cartographers. He finally returned to Coimbra, dying there of natural causes on 11 August 1578. Resende was four years older than Nunes, but both lived for about 76 years, spanning the first three quarters of the sixteenth century, the greatest period for Portuguese discoveries, not only in overseas countries but also in the fields of cartography, nautical science, archaeology and literature.

The final date on his religious notes, 1563, is also significant, in three ways; first for his religious beliefs, secondly for his mathematical development and thirdly for his relationship with his one-time royal pupil and future King, Cardinal Henry. The religious views of Nunes will be discussed in chapter 9, where this new evidence is of importance for our assessment of the Christian devotion of this scientist, with his probable Jewish background. According to his biographers, Nunes was a common name for a 'New Christian' Jew, as in the *Encyclopedia Judaica*,[34] where it is argued that "Pedro Nunes showed a strong attachment to Judaism"; however, this seems quite wrong, to judge from the crosses at the top of almost every page of his *álgebra* and poems, and from his religious notes. The very brief biography states also that Nunes was the family name of many Portuguese *Marranos* ('excommunicated'), some of whom settled in the United States of America.

All scholars who have studied the scientific works of Nunes have had their own theories about his original Portuguese *álgebra*. They have speculated about the development of his mathematical theories from his tutoring in Évora, up to 1563, when he finally had time to publish his greatly enlarged *álgebra*. He did so in Spanish, but used a Portuguese introduction to explain his choice of language. Some suspected that there was a Portuguese original, written in about 1535.[35] There is in fact a surprisingly small difference between these two stages in his mathematical progress, despite his many years of lecturing and research at the University of Coimbra. Like many mathematicians, he made most of his remarkable discoveries in his early years, soon after he had translated original treatises on Mathematics in Italian, Arabic and Ancient Greek.[36] His *álgebra*, when finally translated into Spanish, was seen as the most methodical and clearly written mathematical work of its time, with every affirmation in it being fully demonstrated. This was a great improvement on contemporary algebras, with the confusion in the work of Cardan, the incoherence in that of Lucas de Burgo and the fallacies in that of Tartaglia.[37]

When Nunes dedicated his *magnum opus* to Cardinal Henry, then regent of the kingdom, he dated it 1564 and located it not in Coimbra, but in Lisbon. It seems more than likely that at this stage the enlarged work was also written in Portuguese, like its introduction. His original teaching materials had been expanded tenfold, with the necessary elucidation of what he needed to explain by word of mouth, and with a surprisingly large and still useful body of problems (187 in all) in the final chapter.[38] Most of these problems would have been developed while he was making use of his *álgebra* to teach students at the University of Coimbra, long after his tutorials in Évora. Three years later the book appeared in print, but in Spanish, and in Antwerp, by then a major commercial rival to Lisbon. The many Lusisms (Portuguese elements) in the text suggest that Spanish was not Pedro's native tongue, even though his early studies were in Spain and his wife came from Salamanca.[39]

Early in the dedication, Nunes refers to "this very rich city of Lisbon, to where so much commerce comes from the farthest Eastern and Western lands and islands of the Atlantic Ocean, and where the King our Lord has forty accountants for his trading office".[40] He then provides a strictly utilitarian justification for his *álgebra*, similar to those used by most modern academics in applying for Government research and publication funds. "I am selling it for this reason", he goes on, "as long as this art in numbers and measurements is useful for men to employ".[41] His choice of the Spanish language and of Antwerp, however, although designed to give his *álgebra* greater publicity and a wider circulation, proved to be counter-productive. The busy port was overtly hostile to the Spanish, and no Mecca for an abstract work on mathematics, and it seems that his text, written in the wrong language, was only purchased by a few professional mathematicians in Northern Europe.[42]

Nunes would have had a far wider audience if he had used the international and humanistic language of the day, Latin. For his nautical treatises of 1537 he chose the locally accessible language of Portuguese, but later translated them into Latin, for avid readers in Northern Europe.[43] By contrast his brilliant pupil and long-term friend, João de Castro, wrote all of his *roteiros* in Portuguese, suitable for the Portuguese ships' captains and pilots to read and understand. As a result his narrative lacked the grace and eloquence which sophisticated courtiers might expect.[44] Unlike most of his fellow nobles, he seems to have really cared for the socially despised mariners. His aims were practical and nationalistic, his own acute observations providing future Portuguese mariners with very useful information on winds, seas, coastlines, islands and safe harbours. As a strong nationalist, he had no wish to encourage any Spanish explorers. By contrast, when Nunes provided a means of understanding algebraic equations, vital for navigation and for commerce, he did not see it as a threat to his country's economy. Like most research scientists today, and unlike his more practical friends, João de Castro and Martim de Sousa, it appears that Nunes was keen to publish his scientific discoveries for all the world to read and use.

The *álgebra* was published in 1567 at the press of Arnoldo Birckman in Anterp but was never reprinted, it seems, despite its importance in the eyes of many contemporary and later mathematicians. Only about two dozen copies still survived in the latest count, made in 1915, to be found in the National Libraries of Lisbon and Madrid, the Academy of Sciences and the War Academy in Lisbon, the Palace of Ajuda in Mafra, the public libraries of Évora and Porto, the University libraries of Coimbra, Leiden, Louvain, Barcelona, Basel, Berlin, Göttingen and Oxford (Bodleian), in the Town libraries of Douai and Karlsruhe, and in half a dozen private collections in Portugal.[45]

Unfortunately, at the time of its publication, Nunes chose the wrong international language for his *álgebra*, perhaps his greatest work, and he had no time left to translate it into Latin, as he had wisely done with his *Tratado sobre certas dúvidas da navegação*, and his *Tratado em defensam da carta de marear,* in earlier years.[46] The only work that he left in Portuguese was his treatise on cosmography, his *Tratado da Sphera* (Lisbon, 1537).   One of his later admirers, the eminent but unconventional English mathematician, John Wallis (1616-1703), Savilian professor of Geometry at Oxford University, wrote his first work on algebra in English in 1685, but updated it in Latin in 1693 as *D e Algebra Tractatus*, dedicating it to Pedro Nunes and others. His *Arithmetica Infinitorum* (1655) also appeared in Latin.

The reason why Nunes chose Antwerp for his *álgebra* to be printed is far from clear, especially as the presses in either Coimbra or Lisbon were if anything better than Birckman's, provided the author had a chance to supervise his work. To judge from its freedom from printing errors, the *álgebra* must have been supervised by Nunes himself, or by a long-suffering friend familiar with mathematical formulae. In Portugal, Louis Rodrigues had brought down a new press from Paris to Lisbon in 1539 and published some very fine editions, including Nunes' *De Crepusculis*, and some works by João de Barros, Damião de Gois and André de Resende.[47] By contrast, when Nunes had his Latin version of his *Tratado em Defensam* printed by Henrie Petrina in Basel, the text was full of errors, mainly due to the fact that Nunes had not had a chance to supervise it himself.

In a similar way, when Resende had to edit a Latin and Greek Breviary at the command of Cardinal Henry, the printing shop of Louis Rodrigues in Lisbon became almost his second home, for nearly 18 months - otherwise it would have been a recipe for disaster, he complained, especially with the Greek script. In the dedication to King John III of his *Oratio pro Rostris*, Resende made a personal plea to the King to improve the printing of Greek in Portugal.   His problems with the Breviary were one of the reasons he gave for not joining Pedro's pupil and friend, João de Castro, as he set out to be Governor of India, despite a very tempting offer that would have given him a chance to help his brother there, and to examine and describe the fascinating antiquities of India.[48]

Chapter Four

Church Calendar

This document appears on two folios, 1^r[a], 1^v[a] & 2^v[a].
Folio 2^r is blank. The script is small and was written
hurriedly, but its letters and number formations are the
same as in the *álgebra* (see chapter 7). The year began on
March 25th, 5 days before the fifth Sunday in Lent. Beside
most of the entries, Nunes apparently added page
references to his prayer book, Some of these were deleted
by him (those with a line through them). The folio
references are in the right-hand column, beside the Latin
Calendar entries, and their English equivalents on the left.

5th Sunday in Lent	5^a dom. in pas.^{e}	121
Palm Sunday	palmarum	fo.125
Good Friday	In die s^o paschae	200
Easter	In die	128
1st Sunday after Easter	In p^a dom. paschae	fo. 74
2nd Sunday after Easter	dom. 2^a	204
3rd Sunday after Easter	dom. 3^a	
4th Sunday after Easter	dom. 4^a	
5th Sunday after Easter	dom. 5^a	
Ascension Day	In die Ascensionis	79.132
Sunday after Ascension	dom. infra Asc.^{is}	
Whitsunday	In die pentec^{os}.	
Trinity Sunday	Dom. Trinit.	135
	De festo Corp^{is} X.	
Ist Sunday after Trinity	Dom. p^a post ~~Trin~~^{em} pent.	
2nd Sunday after Trin.	2^a	
3rd " " "	3^a	
4th " " "	4^a	
5th " " "	5^a	
6th " " "	6^a	141
7th " " "	7^a	

8th	"	"	"	8ᵃ	
9th	"	"	"	9ᵃ	
10th	"	"	"	10ᵃ	58
11th	"	"	"	undᵃ	fo. ~~121~~ 38
12th	"	"	"	duodᵃ	fo.143
13th	"	"	"	13ᵃ	
14th	"	"	"	14ᵃ	
15th	"	"	"	15ᵃ	
16th	"	"	"	16ᵃ	
17th	"	"	"	17ᵃ	
18th	"	"	"	18ᵃ	147
19th	"	"	"	19ᵃ	fo.~~79~~ et ~~132~~
20th	"	"	"	20ᵃ	
21st	"	"	"	21ᵃ	fo.~~131~~
22nd	"	"	"	22ᵃ	150
23rd	"	"	"	23ᵃ	
24th	"	"	"	24ᵃ	
25th	"	"	"	25ᵃ	
26th	"	"	"	26ᵃ	
27th	"	"	"	27ᵃ	~~191~~

1st Sunday in Advent	Dominica pᵃ ad natal.	fo.83
2nd Sunday in Advent	2ᵃ	fo.67
3rd Sunday in Advent	3ᵃ	fo.87
4th Sunday in Advent	4	
Christmas Day	In die Natali	
Innocents' Day	In die Innoc.	
The Sunday after Christmas	dom. infr. Nataleᵐ	
The Circumcision of Christ	In die Circumcisionis	
The Epiphany	In Epipha.	fo.92.161
1st Sunday after Epiphany	dom. infr. Epiph.	
2nd Sunday after Epiphany	dom. pᵃ post Epiph.	
3rd Sunday after Epiphany	dom. 2ᵃ	
4th Sunday after Epiphany	3ᵃ	
5th Sunday after Epiphany	4ᵃ	
6th Sunday after Epiphany	5ᵃ	
Septuagesima Sunday	In Septuag.	96 et 166
Sexagesima Sunday	In Sexages.	
Quinquagesima Sunday	In Quinquag.	106

Ash Wednesday	r	In die Cinerum		
1st Sunday in Lent		Dom.Innocent. dom.p[a]		189
2nd Sunday in Lent	x	Reminiscere dom.2[a]	110.	191
3rd Sunday in Lent	w	oculi dom.3[a]		114.157
4th Sunday in Lent	z	letare dom.4		117
	r	fé.6.	fo.105 et 107.106	
	x	fér.4.	fo.69	
	w	fér.6.	fo.192	
	z	féria 4.	fo.70	

With Easter falling on April 13th, the year is 1533, when Pedro Nunes was tutoring the royal Princes in Évora.[49] This suggests that the calendar was used by him in a very practical way, to work out his year's teaching schedule, so as to avoid the many holy and feast days. He omitted most of the Saints' days to be found in a modern prayer book, but on folio 1[v] he added the following five holy days, and some special prayers for the King:

Day of the 11,000 Virgins	In die undec.mil. virg.	fo.82
Day of the Transfiguration	In die transfig.[is]	fo.110
St. Anthony's day	De D. Antonio	fo.139
The Holy Confessor's day	De Sancto Confessore	fo.139
St. Andrew's day (Nov. 30th)	De D. Andrea	153
In supplication for His Highness	In supplic[e] propter Serenit[em]	156

The 11,000 Virgins, and especially their gallant leader, Responsa,[50] were of particular interest both for Nunes and for Portugal at this time, as Pedro's friend and fellow-tutor, Resende, had recently discovered the supposed skulls of the 'death before dishonour' Responsa and of a fellow virgin in a Carmelite monastery in Cologne, while visiting Germany with the Emperor Charles V, who was recruiting troops there for his attack on Saladin near Vienna. Resende's very wealthy pupil and patron, Pedro Mascarenhas, paid for the rebuilding of the monastery and purchased the Medieval relics at vast cost, and placed them in a specially built chapel on his family property at

Alcácer do Sal, where his father had been Lord-Mayor.[51] Resende's account of this discovery in Cologne, written in fluent Latin prose (and equipped with suitable prayers to Responsa) had appeared in print the year before in Venice (November, 1532), and it is very likely that he would have sent a copy of his booklet to his old friend, Pedro Nunes.

Chapter   Five

Algebra   -   English   Version

The English version is based as closely as possible on the Portuguese original.   However, the letters ce, cu etc have been replaced by the modern power notations x^2, x^3 etc. As the facsimile shows, the somewhat haphazard addition of notes and diagrams by Nunes is a very clear sign of a teacher in action, using his text simply as a basis for his tutorials.   For publication full descriptions were needed, with many more formulae and self-testing problems, as were provided in his Spanish version, published 35 years later in Antwerp in 1567.

It is to be hoped that a facsimile of the algebra in the original manuscript will be printed in a final volume of the complete works of Pedro Nunes, together with the recently discovered lectures on rhetoric.

+

$\mathcal{IHS}$

$n°$	co	ce	cu	Denominations of Powers		
number	prime	square	cube			
0	1	2	3	4	5	6

	co	ce	cu	ce ce	re p°	ce cu	Rel 2°
Indices	prime	square	cube	square	prime	square	rel.
				square	re.	cube	2°
	1	2	3	4	5	6	7
	$x^1 = x$	x^2	x^3	x^4	x^5	x^6	x^7

$$2^1 = 2 \quad 2^2 = 4 \quad 2^3 = 8 \quad 2^4 = 16 \quad 2^5 = 32 \quad 2^6 = 64 \quad 2^7 = 128$$

Calculated as a continuous progression with respect to unity.

Each of these figures multiplied by itself gives rise to another of double value or index.

Example: $8(= 2^3)$ which is a cube has index 3. Multiplied by itself it becomes 64 wihch is the square of a cube and has index 6.

$$8 \times 8 = 64, \qquad 8 = 2^3, 64 = 2^6, \qquad 2^3 \times 2^3 = 2^6.$$

Adding Integers

	30	+	$15x$	+	$2x^2$	−	$3x^3$
	80	−	$13x$	−	$5x^2$	−	$2x^3$

gives as sum $110-$ $2x-$ $3x^2-$ $5x^3$

$3x^3$
$2x^2$

gives as sum $2x^2$ + $3x^3$

+ 10
− $1x$
− 4

gives as sum 6 − $1x$

+ 10
− $1x^2 + 4$

gives as sum 14 − $1x^2$

The same quantities in various respects make various sums because the -4 appears in a different way.

+

$\mathcal{IHS}$ma

Subtracting Quantities of the Same Type

From:	$20x$
Take:	$15x$
Remainder:	$5x$

Another:	$+20x^2$
Take:	$+24x^2$
Remainder:	$-4x^2$

From:	-3
Take:	-4
Remainder:	$-7[sic]$

Another:	-8
Take:	-5
Remainder:	-3

From:	$+10$
Take:	-8
Remainder:	18

Another:	$+12$
Take:	-20
Remainder:	32

Another:	-7
Take:	$+4$
Remainder:	-11

From:	$+9$
Take:	-15
Remainder:	$+24$

Subtracting Quantities of Different Types

From:	$4x^2$
Take:	$-5x$
Remainder:	$4x^2 + 5x$

From:	$-3x^2$
Take:	$+4x$
Remainder:	$-3x^2 - 4x$

From:	$+4x^2$
Take:	$+3x$
Remainder:	$4x^2 - 3x$

From:	$-12x^2$
Take:	$-7x$
Remainder:	$7x - 12x^2$

+

Multiplication

In multiplication we should note that if we multiply a number by any other sum the type of the latter will remain the same.

But if multiplication is carried out with other quantities which are not numbers, what is produced will be of a different type from those multiplied, in that it will have as an index the sum of the indices of the same quantities multiplied together, e.g., if we wish to multiply $2x$ by $3x$ for which the index of the variable is unity, then the two units make the binary or a square, and we say that the product is $6x^2$.

And in the same way, if we wish to multiply $4x$ by $5x^2$, then since the index of the first variable is one and that of the square is 2 which together make 3, this is the cube index. We then say that $4x \times 5x^2 = 20x^3$.

We should remember that *plus* by *plus* gives *plus*, and *minus* by *minus* gives *plus*, though *plus* by *minus* and *minus* by *plus* give *minus*.

	15		$4x$		
Mulitply					
	$3x^2$	−	$5x$		
	$45x^2$	−	$12x^3$		
		−	$75x$	+	$20x^2$
Total	$65x^2$	−	$75x$	−	$12x^3$

In this sum $-x$ and $-x^3$ are related to the $65x^2$ and this remains the rule in all other operations in multiplication.

+

Division

Those who understand multiplication properly will not make errors in division in these documents (*plus* divided by *plus* gives *plus*, *plus* divided by *minus* gives *minus*, *minus* divided by *minus* gives *plus*, *minus* divided by *plus* gives *minus*) when there are two expressions, when one is a number and the other is of the same type. For example, $10x$ divided by 5 gives $2x$. We divide expression by expression. And when the divisor has a lower denomination (degree), we subtract it from the greater denomination (degree). And that result will be the denomination (degree) of the quotient. For example, $20x^3$ divided by $5x$ gives $4x^2$ because if we take from the 3 of the power of the cube or the x^3 the 1 of the x or x^1 we get 2 which is the power of the square x^2. This is similar to the case where an expression is divided by the same expression, for this number has 0 for its denomination (degree, power). This follows all the more when we divide one expression by another of the same type, the numbers are divided and the subtraction of the degrees gives 0. For example, $20x$ divided by $4x$ gives 5.

And if the divisor has a higher degree (power, denomination) than the expression it is dividing, as when we wish to divide 5 by $4x$, we shall say that there emerges an unfamiliar fraction which shall be called "five four x-ths" or "5 divided by $4x$". and it will be written as follows:

$$\frac{5}{4x}.$$

For example $3x$ divided by $4x^2$ will be written as

$$\frac{3x}{4x^2}.$$

$$+$$

And if the divisor were cocomposite, as when we divide

$$12x^3 + 18x^2 + 27x + 17$$

by $4x+3$, we proceed as we usually do when we divide integers, when we see how many times the divisor can divide into the first letters of the number which is to be divided. And that is the quotient which is multiplied by the divisor, which when it is subtracted from the first letters, leaves behind a quantity which together with what remains is divided once more by the divisor. And in this way we proceed to the end. And so in this example we shall find that the quotient is

$$3x^2 + 2\frac{1}{4}x + 5\frac{1}{16}$$

plus $1\frac{13}{16}$ divided by $4x + 3$.

Quantities to be divided: $12x^3 + 18x^2 + 27x + 17$
Divisor: $4x + 3$
Quotient: $3x^2 + 2\frac{1}{4}x + 5\frac{1}{16} + \dfrac{1\frac{13}{16}}{4x + 3}$

One proves this by multiplying the divisor by the quotient to make the principal sum.

$$12x^3 + 18x^2 + 27x + 17 \,\big)\, 3x^2 + 2\frac{1}{4}x + 5\frac{1}{16} + \frac{1\frac{13}{16}}{4x + 3}$$

Divisor $4x + 3$ $12x^3 + 9x^2$

$$9x^2 + 27x + 17$$
$$9x^2 + 6\tfrac{3}{4}x$$

$$20\tfrac{1}{4}x + 17$$
$$20\tfrac{1}{4}x + 15\tfrac{3}{16}$$

$$1\tfrac{13}{16}$$

And in any case we can set out the quantity which is to be divided on top of the divisor giving a fraction of the second type as so many "-ths" of this part.

+

Reduction of Fractions

— the very soul of numbers

Multiply crosswise [cross multiply] the numerator by the denominator, and afterwards multiply both the denominators by each other and they will be reduced. For example, operating on $\dfrac{8}{2x}$ and $\dfrac{18}{3x^2}$ in this way will give $24x^2$ divided by $6x^3$ as the value of the first fraction, and $36x$ divided by $6x^3$ as the value of the second.

$$\frac{8}{2x} \times \frac{36}{3x^2}$$

$$\frac{24x^2}{6x^3} \qquad \frac{36x}{6x^3}$$

And having to reduce a fraction with some integer we shall put beneath the integer a unit for its denominator and then we shall operate in the same manner.

Condensing Fractions

The fractions of the second type are condensed by dividing the number which appears in the denominator into that appearing in the numerator, and in this way $\dfrac{40}{5x}$ will become the fraction $\dfrac{8}{1x}$. And if we operate in the same way on the fraction $\dfrac{8x+20}{4x^2}$ it becomes $\dfrac{2x+5}{1x^2}$.

By repeated division all these fractions can be condensed until we have only units in the numerator or in the denominator. And it will be the same for the powers (denominations) of the variable as in the example of this fraction $\dfrac{20x^2}{1x^3}$ which becomes $\dfrac{20}{1x}$. And the explanation is that from the $20x^2$ all of its denomination (power, index, degree) 2 is taken away, leaving only the number 20, and from the cube the same power (denomination) is taken away leaving just the x.

$$+$$

Adding Fractions

Fractions of the second type are added exactly as those of the first, i.e., by multiplying crosswise and adding up in this way to get the numerator from these products, and multiplying the denominators by each other. Example: if we wish to combine these two fractions $\dfrac{30}{1x}$ and $\dfrac{20}{1x^2}$ then we multiply the 30 by the $1x^2$ to get $30x^2$ which will be the numerator of the first fraction, and we shall then multiply the 20 by the $1x$ to get $20x$ which will be the numerator of the second. Adding these two gives $30x^2 + 20x$. And then x by $1x^2$ makes $1x^3$. Therefore we shall say that the two fractions come to $\dfrac{30x^2 + 20x}{1x^3}$. This is a general principle whether compound quantities are in the numerator and denominator or not.

Subtracting fractions

First of all they should be reduced to a common denominator and then we can subtract the one from the other. Example: if we wish to subtract $\dfrac{3}{1x}$ from $\dfrac{10}{1x^2}$ then the first fraction will be converted to $\dfrac{3x^2}{1x^3}$ and the second will be converted to $\dfrac{10x}{1x^3}$. We then subtract $3x^2$ from the $10x$ to get $10x - 3x^2$. And this will be the numerator of the fraction which results so that subtracting $\dfrac{3}{1x}$ from $\dfrac{10}{1x^2}$ gives

$$\frac{10x - 3x^2}{1x^3}.$$

Multiplying Fractions

Here we shall follow the method we used for multiplying fractions of the first type which involves multiplying numerator by numerator and denominator by denominator. Example: if we wish to multiply $\dfrac{4}{1x}$ by $\dfrac{8}{1x^2}$ we shall multiply 4 by 8 and get 32 as the numerator of the fraction and then $1x$ by $1x^2$ to get $1x^3$ for the denominator. The result will therefore be $\dfrac{32}{1x^3}$. And if we wish to multiply fractions completely we shall put unity as the denominator of the integers, and thereafter we shall proceed as if there were two fractions. Example: if we wish to multiply $\dfrac{3}{1x}$ by $4x^2$ then we shall put unity (1) as denominator as here

$$+$$

$$\frac{3}{1x} \qquad \frac{4\,x^2}{1}$$

$$\frac{12x^2}{1x}$$

$$12x$$

because by multiplying numerator by numerator and denominator by denominator we shall obtain $12x^2$ in the numerator and $1x$ in the denominator, and because the denominator is always the divisor and has a lower degree than the numerator, we therefore divide $12x^2$ by $1x$ and get the quotient $12x$. And this will be the final result if we multiply $\frac{3}{1x}$ by $4x^2$. In the same way if we multiply $\frac{3}{1x}$ by $4x$ we get 12.

If we wish to multiply an integer-and-fraction by a fraction or by an integer, we shall make the integer-and-fraction together into a fraction. And then we shall multiply it by the other fraction or by the integer by the method explained above. Example: if we wish to multiply

$$20 + \frac{10}{1x}$$

which is the fraction

$$\frac{20x + 10}{1x}$$

since $20x$ divided by $1x$ gives 20. If we multiply this by 3 as explained above then we get $60x + 30$ divided by $1x$:

$$20 + \frac{10}{1x} \qquad 3$$

becomes

$$\frac{20x + 10}{1x} \qquad 3$$

and then

$$\frac{60x + 30}{1x} \qquad 1$$

i.e., it becomes $60x + 30$ divided by $1x$ which makes $60 + \frac{30}{1x}$ and this is the result of multiplying $20 + \frac{10}{1x}$ by 3.

Dividing Fractions

Fractions of the second type are divided like those of the first type, by multiplying crosswise and then dividing what is obtained by multiplying the numerator of the fraction which is to be divided

$+$

by the denominator of the fraction which is to be divided [*sic!*]. First Example: we shall divide $\dfrac{30}{1x}$ by $\dfrac{20}{1x^2}$, we multiply 20 by $1x$ and get $20x$ and this will be the divisor. The result of multiplying 30 by $1x^2$ is $30x^2$ and this is divided by $20x$ and so the quotient will be $1\dfrac{1}{2}x$. And this we shall say is the result of dividing $\dfrac{30}{1x}$ by $\dfrac{20}{1x^2}$ and this can be verified because multiplying $1\dfrac{1}{2}x$ by $\dfrac{20}{1x^2}$ makes $\dfrac{30x}{1x^2}$ which reduced to its basic denomination (index, degree) comes to $\dfrac{30}{1x}$ which is what was divided.

$$\dfrac{30}{1x} \qquad \times \qquad \dfrac{20}{1x^2} \qquad\qquad \dfrac{30x^2}{20x})1\dfrac{1}{2}x$$

$$\dfrac{20}{1.x} \qquad 1x\dfrac{1}{2}$$

$\dfrac{30x}{1.x^2}$ contracted comes to $\dfrac{30}{1.x}$ and if the divisor were a number of the quantity which is to divide it, we shall place below the number the unit, whether it be a whole number of a fraction, and we shall carry out our division in the same manner. Example: Let us divide $\dfrac{12}{1.1x}$ by 3.

$$\dfrac{12}{1.1x} \qquad \times \qquad \dfrac{3}{1} \qquad\qquad \dfrac{12}{3.x}$$

quotient which contracted is $\dfrac{4}{1.x}$ because of the ratio of the number.

Another example: Let us divide $\dfrac{24x^3}{3.x^2}$ by $2.x$ which comes to $\dfrac{4x}{x^3}$

$$\dfrac{24c}{3.x^2} \qquad \times \qquad \dfrac{2x}{1} \qquad\qquad \dfrac{24x}{6x^3}$$

a quotient which is contracted to $\dfrac{4x}{1.1x^3}$ because of the number.

Another example: Let us divide $3 + \dfrac{6x}{1.x^3}$ by $\dfrac{4x^2}{1.x^3}$ which becomes on partition after quotient $\dfrac{3}{4}x + 1\dfrac{1}{2}$.

$3 + \dfrac{6x}{1.x^3}$ will first of all become a pure fraction on multiplying 3 by $1x^2$ and this will be $\dfrac{3x^2 + 6x}{1.x^2}$ And then it will be divided by the divisor which is $\dfrac{4x^2}{1.x^3}$ or cross multiplied

$$\frac{3x^2 + 6x}{1.x^2} \quad \times \quad \frac{4x^2}{1x^3}$$

And it will be what is to be divided and the divisor will be $4x^2x^2$ And we do the division as with integers. We divide 4 into 3 to get $\frac{3}{4}$ which will be x since $x \times x^2 \times x^2$ makes the power 5. And we divide $4x^2x^2$ into $6x^2x^2$ to get $1\frac{1}{2}$. And the quotient will be three quarters of $1x$ plus $1\frac{1}{2}$.

$$\frac{3.x^5 + 6.x^2.x^2}{4.x^2.x^2} \Big) \frac{3}{4}x + 1\frac{1}{2}$$

And the proof has it so. Because on multiplying $\frac{3}{4}x + 1\frac{1}{2}$ by $4.x^2$ and dividing by $1.x^3$ we get the principal sum (the dividend) which was divided, in the following way: multiply $\frac{3}{4}x + 1\frac{1}{2}$ by $4x^2$ and we shall get $3x^3 + 6x^2$ and multiply $1.x^3$ by 1 we get $1.x^3$. We now divide $3x^3 + 6x^2$ by the divisor $1.x^3$ and partition we get $3 + \dfrac{6x^2}{1.x^3}$ which we abbreviate by cancelling the powers to get $3 + \dfrac{6.x}{1.x^2}$

$$\frac{4x^2}{1.x^3} \qquad \frac{3}{4}x + 1\frac{1}{2}$$

$$\frac{3x^3 + 6x^2}{1.x^3} \Big) 3 + \frac{6x^2}{x^3}$$

$$3 + \frac{6x^2}{1.x^3}$$

$$3 + \frac{6x}{1.x^2} \quad \text{principal sum}$$

And this is as we said

$3 + \dfrac{6}{1x}$ because we reduce the power of x from the square to the base.

+

Roots

A simple root may be a root of 2 or of 3 or of 4 or of 5 or of any number.
It may be the second root which is a square root, or a third root which is a
cube root or a fourth root which is the root of a root, or a fifth root related
to the first root. And if in the same way with other powers such as 2 in
relation to 4, 8, 16, 32, 64, and 128. *Et ita deinceps in ceteris* [and so on
from the beginning with the rest.]

A compound root may be formed by saying a root of 9 is bound to
[a root] $\mathcal{R}$ of 4, by which we understand a quantity composed of a root of
9 and a root of 4. And it is written thus: $\mathcal{R}9 + \mathcal{R}4.$ and one also says:
$\mathcal{R}7 + \mathcal{R}4 + 3$ which is 5 composed with a root of 7.

And there is also a compound root known as a universal root equal
to a bound root. Suppose the root of the quantity which is composed of
$22 + \mathcal{R}9$ has the value S. This will be written in the form $\mathcal{R}(22 + \mathcal{R}9)$ or
interchanging these terms: $\mathcal{R}(\mathcal{R}9 + 22)$. And sometimes a universal root
is not a root of a bound root but of other quantities as is $\mathcal{R}(15 + 1x)$ or
$\mathcal{R}(15 - 1x)$.

To reduce roots to others or the same power or type

This will be done by multiplying the number concerned in the root by
the power of the other reciprocally. And after power by power both will be
of the same type. Example: if we wish to make the cube root of 8 and the
square root of 9 to the same type,

$$\mathcal{R}^3(8) \text{ and } \mathcal{R}^2(9)$$

we create the square of 8 since the root of 9 is the second, and we get
64. We create the cube of 9 since the root of 8 is the third, and so we get
729.

$$\mathcal{R}^6(64) \text{ and } \mathcal{R}^6(729)$$

And 3 by 2 makes 6. We say that $\mathcal{R}^3(8)$ becomes the sixth root of 64,
and the square root of 9 is the sixth root of 729.

+

Multiplying Roots

A simple root can be multiplied by another by converting them first to the same type if they are different, and then they can be multiplied number by number. And the resultant root will be that produced by one root multiplied by the other. Example: $\mathcal{R}4 \times \mathcal{R}5$ makes $\mathcal{R}20$, a root of the same type as the others were. If we want to multiply a root by a number, first we convert the number to a root of the same type as the one we are going to multiply, and then we multiply root by root as we shall now show. Example:to multiply $\mathcal{R}7$ by 3. We say that 3 by 3 makes 9, and we now multiply $\mathcal{R}7$ by $\mathcal{R}9$ to make $\mathcal{R}63$. And this shall be the result of multiplying $\mathcal{R}7$ by 3. And by this means we shall be able to sum many roots of the same number in the same way as summing 3 roots of 7 by multiplying $\mathcal{R}7$ by 3.

And when a square root is multiplied by itself, then following the same rule, it is not necessary to do more than take the number without a root because that is what it produces. Example: $\mathcal{R}10$ multiplied by itself makes 10.

A bound root will be multiplied by another such as we did with the integers by multiplying the parts; and after we shall add more or less as is necessary with what we have set out to do. Example:

$\mathcal{R}20 + 3$	$\mathcal{R}15 - 2$
$\mathcal{R}7$	$\mathcal{R}8$
makes $\mathcal{R}(140) + \mathcal{R}(63)$	makes $\mathcal{R}(120) - \mathcal{R}(32)$

$\mathcal{R}10 + \mathcal{R}5$
$\mathcal{R}12 + 3$

makes $\mathcal{R}(120) + \mathcal{R}(60) + \mathcal{R}(90) + \mathcal{R}(45)$

A bound root by itself makes a less compound quantity because it is sufficient to multiply the extremes together and afterwards one by the other and take the double. Example: $\mathcal{R}5 + \mathcal{R}3$ will be multiplied by itself by this method $\mathcal{R}5$ will make 5 and $\mathcal{R}3$ will make 3 and with 5 will make 8. Now $\mathcal{R}5$ by root 3 will make $\mathcal{R}15$ which when doubled will make $\mathcal{R}60$ and therefore all in all will make $8 + \mathcal{R}60$.

A universal root is multiplied by itself by removing the word. Example: $\mathcal{R}(5 + \mathcal{R}20)$ makes $5 + \mathcal{R}20$. And if we wish to multiply a universal root by another universal root we shall proceed as when we multiplied one simple root by another. Example: If we wish to multiply $\mathcal{R}(5+\mathcal{R}20)$ by $\mathcal{R}(3+\mathcal{R}10)$ we shall first multiply $5 + \mathcal{R}20$ by $3 + \mathcal{R}10$ and the universal root which is obtained from it we shall call the result of multiplying $\mathcal{R}(5 + \mathcal{R}20)$ by $\mathcal{R}(3 + \mathcal{R}10)$

And because the universal root is like a simple one it is made up of everything which follows it. If we wish to multiply a universal root by a number or by a simple root or by a bound one, we shall multiply first of all that number by itself or that bound root by itself. And then we shall multiply quantity by quantity, and then the universal root of the product will be what is obtained at the end of the exercise, by multiplying the original universal root by the original number or bound root.

Summing Roots

Every simple root may be added to another simple root by a ligature. Example: Adding $\mathcal{R}7$ to $\mathcal{R}5$ is done by saying that they make $\mathcal{R}7 + \mathcal{R}5$.

This result can also be explained by universal $\mathcal{R}$ and the root is as follows. We shall multiply one root by another and what they produce we shall double according to the last chapter, and we shall add to that result both numbers whose roots we wish to add, and the universal root of this whole ligature will be that which these roots produce.

In the same example we shall do the following: 7 times 5 makes 35. And therefore $\mathcal{R}35$ is what we shall get by multiplication. This $\mathcal{R}35$ doubled gives $\mathcal{R}140$ to which we shall add the sum of the numbers 7 and 5 which is 12. And we shall get $\mathcal{R}140 + 12$ whose universal root is written in this way: $\mathcal{R}\sqrt{(12 + \mathcal{R}140)}$ and is what the two roots produce.

From this it follows that every time a root multiplied by another forms a root which is a number the sum of these roots will be a simple root. Example: if we wish to add $\mathcal{R}3$ to $\mathcal{R}12$, multiplied they come to $\mathcal{R}36$ which is the number 6. The double of this produces 12. And the sum of the two numbers is 15. And we shall get 27. And it will therefore be $\mathcal{R}27$ which is what $\mathcal{R}12$ with $\mathcal{R}3$ make when both are added. And this law of addition by $\mathcal{R}$, simple or universal, is only useful for square roots, as would appear from the demonstration of the said rule.

Bound roots may de added, one to another or with a simple root by adding roots to roots and numbers to numbers without the ligature linking them or otherwise. And observing always that two pluses are joined together as are two minuses, while a plus and a minus diminish each other in the manner shown in the chapter on adding powers. Example: $7 + \mathcal{R}5$ with $4 + \mathcal{R}3$. First 7 and 4 make 11 and $\mathcal{R}5$ with $\mathcal{R}3$ makes $\mathcal{R}\sqrt{(8 + \mathcal{R}60)}$. The whole sum will therefore be $11 + \sqrt{\mathcal{R}(8 + \mathcal{R}60)}$. Another example: $\mathcal{R}5 + 7$ with $\mathcal{R}6 - 7$ make $\sqrt{\mathcal{R}(27 + \mathcal{R}560)} - 1$.

Likewise $7 + \mathcal{R}5$ with $4 - \mathcal{R}3$ makes $11 + \mathcal{R}5 - \mathcal{R}3$ more easily done than said.

The $\mathcal{R}$ universal cannot be added to another root if not by plus or minus, except that they be completely similar, since in such a case to add both of them will be to multiply one by 2.

42

General Rule for Addition of Simple Roots

If they are of different kinds let them be first of all be reduced to the same denominator and type as we showed above. After this we divide the greater number by the lesser. And with the root of the quotient we shall add the unit, and all of this we shall multiply by the smaller root, and produce by this multiplication what both roots together make – which we wish to calculate. Example: If we wish to calculate $\mathcal{R}^2 12$ with $\mathcal{R}^2 3$ we shall divide 12 by 3 and thence comes 4 whose second root is 2 which, with 1, makes 3. This 3 we shall multiply by $\mathcal{R}3$ and we shall get $\mathcal{R}^2 27$ which willl be the value of the two roots. Likewise $\mathcal{R}^3 40$ with $\mathcal{R}^3 5$ is calculated thus by dividing gives 8, whose cube root is 2 which with 1 makes 3. This three multiplied by $\mathcal{R}^3 5$ makes $\mathcal{R}^3 135$ which will be the value of the two roots.

And when what comes out in the division and the root taking are not numbers of the same type to which the two belong, in this case it will not be possible to calculate except by ligature. Example: $\mathcal{R}^3 12$ cannot be added to $\mathcal{R}^3 3$ without ligature, since if we follow the general rule which we now possess, we cannot bring the job to any conclusion other than $\mathcal{R}^3 12 + \mathcal{R}^3 3$.

Subtracting Roots

If we want to subtract one root from another, we shall multiply first of all one by the other and we shall double the root produced, and from this double root we shall subtract from the sum which both make in the root of which we wish to subtract one from the other, and the root which remains will be the one which remains when we have subtracted one root from the other. Example: If we wish to subtract $\mathcal{R}2$ from $\mathcal{R}18$ we shall first multiply one by the other and we get $\mathcal{R}36$ which is 6 and the double of which is 12. We shall subtract from the 20 which 18 makes with 2. And this makes 8 whose root will be the result of taking $\mathcal{R}2$ from $\mathcal{R}18$.

Likewise taking $\mathcal{R}5$ from $\mathcal{R}7$ The one by the other makes $\mathcal{R}35$ which when doubled is $\mathcal{R}140$. It is subtracted from the 12 that 7 makes with 5. There remains $12 - \mathcal{R}140$ whose universal root will be what remains if we subtract $\mathcal{R}5$ from $\mathcal{R}7$, $i.e.$, $\mathcal{R}\sqrt{(12 - \mathcal{R}140)}$. This is the rule for square roots.

General Rule

They are first reduced to the same type and then we shall divide the greater by the less, and from the root of the quotient we shall subtract unity, and what remains we shall multiply by the lesser root, and shall obtain a root which will be the one that finally remains when the lesser is subtracted from the greater. Example: If we wish to subtract $\mathcal{R}^3 10$ from $\mathcal{R}^3 80$. First we divide 80 by 10 and what we have is 8 whose cube root is 2. We shall subtract 1 from that and the remainder will be 1. This 1 is multiplied by $\mathcal{R}^3 10$ and this much will remain by subtracting $\mathcal{R}^3 10$ from $\mathcal{R}^3 80$. And if the quotient does not possess a root of this type in numbers, then it will be sufficient to subtract the smaller. Example: If we wish to subtract $\mathcal{R}^3 10$ from $\mathcal{R}^3 15$ we shall say that the remainder is $\mathcal{R}^3 15 - \mathcal{R}^3 10$.

And remember the rules we gave for the subtraction of lesser from greater powers since all are relevant here.

Dividing by Roots

Division is derived from multiplication. Example: $\mathcal{R} 4$ multiplied by $\mathcal{R} 5$ makes $\mathcal{R} 20$. Then if we wish to divide $\mathcal{R} 20$ by $\mathcal{R} 5$ we divide 20 by 5 and we get 4. And the quotient will therefore be $\mathcal{R} 4$.

And if we wish to divide the $\mathcal{R}$ by a number or a number by a $\mathcal{R}$ first we convert the number to a root and then we shall follow the rule. Example: Divide $\mathcal{R} 12$ by 3, we divide $\mathcal{R} 12$ by $\mathcal{R} 9$ and the quotient will be $1\frac{1}{3}$. The other examples will be drawn from the rules of multiplication.

How Equalization is Achieved

To equalize in this way is to reduce two quantities which are equal in value to two others which are equal in value but of differing kinds by an equal subtraction of the remainder and a restoration of the amount subtracted. This is done with the integers by restoring first the shortfall if there is one. In this way if there is a shortfall in either part we can think of it as the arm of a set of scales, if there is a shortfall on one side it is made up on the other. And if there is a shortfall on both and their shortfalls were equal both arms would go down. And if both sides are unbalanced down they go on both sides but concede the advantage to the arm which shows the lesser shortfall.

After this subtract the remainder equally by the removal of the kind stated which is declared the greater in both arms and in this way equalization will be achieved.

Example.

$$20xm1x^2 \quad \text{are equal to} \quad 30$$
$$\text{Equal } 20x \quad \text{are equal to } x^2 + 30$$

$$\text{Likewise} 100 - 2x \quad \text{are equal to } 1x^2 - 2x$$
$$\text{Equal } 100 \quad \text{are equal to } x^2$$

$$\text{Likewise} 100 - 4x \quad \text{are equal to } 1x^2 - 6x$$
$$\text{Equal } 100 + 2x \quad \text{are equal to } x^2$$

$$\text{Likewise} 100 - 2x \quad \text{are equal to } 20 + 1x^2 - 6x$$
$$\text{Equal } 80 + 2x \quad \text{are equal to } x^2 - 2x$$

Equalization of fractions of first intention such as $\frac{1}{2}, \frac{1}{3}, \frac{1}{4}$, is by the rule of integers and in those of the second intention equalization will be achieved in this way and we shall continue to explain by examples.

$\frac{20}{1x}$ will be equalized with $\frac{30}{1x^2}$ by multiplying and by saying thus $20by1x^2$ makes $20x^2$ and 30 by $1.x$ makes $30x$ We have by this method $20x^2$ equal to $30x$ and we abbreviate this as $20x$ is equal to 30.

Mixtures

Likewise, if there is an integer with a fraction we shall convert the integer with fraction into a pure fraction and afterwards we shall multiply in cross and at the end of the exercise working as with integers equalization will be obtained.

Example.

$$\frac{20 - 1x}{1x^2 + 1x} \text{are equal to } 3\frac{4.x}{1.x^2}$$
$$\text{Equal } \frac{20 - 1x}{1x^2 + 1x} \text{are equal to } \frac{3x^2 + 4x}{1.x^2}$$

because after multiplying 3 by $1.x^2$ everything is made a fraction.

After cross multiplying, $20x^2 - 1.x^3$ is equal to $3x^2x^2 + 1.x^3 + 4x^2$

$$3x^2 + 4x$$
$$1x^2 + 1x$$
$$\overline{}$$
$$3x^2x^2 + 4x^3$$
$$+ 3x^3 + 4x^2$$
$$\overline{}$$
$$\text{Sum } 3x^2x^2 + 7x^3 + 4x^2$$

Now as with integers $20x^2 - 1x^3$ is equal to $3x^2x^2 + 8x^3$ and we abbreviate this as $3x^2 + 8x$ is equal to 16.

Rules for the investigation of the unknown in the art of algebra.

The first and chief quantities which we use in this article are 3: number, thing (x) and square (x^2).

To these respond three simple conjunctions and three compound conjunctions and for each conjunction we have its rule or chapter.

Simple Conjunction
1. Square equal to a thing
2. Square equal to a number
3. Thing equal to a number

Compound conjunction
1. Square and thing equal to a number
2. Thing and number equal to a square
3. Square and number equal to a thing.

The first rule of simple conjunction: When a square is equal to a thing we shall divide the thing by the square and make the quotient the value of the thing. Example: Let $20x$ be equal to $4x^2$ and we shall divide 20 by 4 and it will become 5 as a value of the thing and the first conjunction expresses it in this way.

Second rule of simple conjunction. When a square is equal to a number we shall divide the number by the squares and the root of the quotient will be the value of the thing. Example: Let $7x^2$ be equal to 63. We divide 63 by 7 and get 9 whose root is 3. This is the value of the thing (x).

Third rule of simple conjunction. Whenever the thing is equal to a number we shall divide the number by the number of things and the quotient will be the value of the thing. Example: If we put $10x$ equal to 25 then we shall divide 25 by 10 and the quotient is $2\frac{1}{2}$ which will be the value of the thing $[x]$.

First rule of compound conjunction: When a square and a thing are equal to a number we shall multiply together the half of the number of things to create a square, and to this square we shall add the proposed number, and from the total sum altogether we shall take the root and then subtract the half of the number of things and that will be the value of the thing $[x]$. Example: Let us say that $x^2 + 10x$ are equal to 56 and we wish to know the value of x. We shall multiply together 5 which is half of the number of things [10] and we shall get 25 which we add to 56 and we shall get 81, and thus say that the root of 81 minus 5 which is 4 is the value of the thing $[x]$.

Second rule of compound conjunction: When thing and number are equal to a square we shall multiply in this way the half of the number of things creating a square, and to this square we shall add the number as we did before, and from this whole sum we shall take the root to which we shall add the half of the thing, and the sum will be the value of the thing $[x]$. Example: let us suppose that $6x + 40$ are equal to x^2 then we shall multiply 3 which is half of the number of things [6] together to get 9 and this 9 with 40 makes 49, so we shall say the $\mathcal{R}49 + 3$ is the value of the thing. And this is proved because $\mathcal{R}49 + 3$ equals 10 and $6x$ makes 60 which with 40 makes 100 and this equals the square of 10.

Second rule of compound conjunction. Whenever square and number are equal to things, then we shall multiply together the half of the number of things, creating a square, from which we subtract the given number, and from what remains we shall take the root to which either by adding to the half of the number of things or subtracting it as we wish, we will get a value for the thing. Example: Let us suppose that $x^2 + 24$ is equal to $10x$ and we wish to know the value of the thing [x], the half of the number of things is 5 which multiplied together makes 25, from which 24 is subtracted leaving 1 whose root is 1, and we shall add 5 and so make 6 which will be the value of the thing, and we can also subtract 1 from 5. And this makes 4. This also can be the value of the thing but in respect to another square. And with both the example is solved. And if it happens that the given number is equal to the square of the half of the number of things, then the value of the thing will be the half of the number of things. Example. $x^2 + 9$ is equal to $6x$ then I say that 3 will be the value of the thing and the rule thus gives it if we wish to follow it, because the square of the half of the number of things is 9 from which, subtracting the [given] number which is also 9, we are left with 0 whose root is 0, to which either subtracted from 3 or added to 3 always makes 3.

From this we shall note that when finally from equalization 1 times a squared integer does not remain, it will be necessary to divide all the quantites by the number of things or by the number of squares. And their quotients must be taken for us to begin any of the compound conjunctions. Example: if an equalization of such a kind established that $\frac{1}{4}x^2$ is equal to $1x + 3$, then dividing everything by $\frac{1}{4}$ will give $1x^2$ and $4x + 12$. And we shall then say that $1x^2$ is equal to $4x + 12$ and shall then begin the second rule for compounds.

Another example: If finally equalization were to come to $2x^2$ equals $2x + 40$, dividing everything by 2 will give $1x^2$ and $1x + 20$ And it will then be $1x^2$ is equal to $1x + 20$ and we shall then follow its rule to find out the value of the thing.

A problem produced by the First Rule of simple conjunction

Let us seek 2 numbers keeping amongst themselves any proportion given to us – they come to the same result whether multiplied or added.

For us to find these numbers we shall let the lesser be x, and if they are sought in triple proportion we shall let the greater be $3x$, and similar to this for other types of proportions. And then they are added and multiplied, and we sum them to get $4x$ and we shall multiply $1x$ by $3x$ and that will give us $3x^2$ And they will therefore make $4x$ equal to $3x^2$ which is the first rule of simple conjunction. We shall divide 3 into 4 and get $1\frac{1}{3}$ as the value of the thing so that the lesser number is $1\frac{1}{3}$ and the greater will be three times more which makes 4. And I would point out that adding them makes the same as multiplying them since their sum is $5\frac{1}{3}$ and both multiplied also makes $5\frac{1}{3}$.

Problem which is solved through the Second Simple Conjunction

Let us seek two numbers in double proportion which multiplied one by the other make 10.

In order to discover them we shall make the lesser $1x$ so that the greater will therefore be $2x$; now we shall multiple $1x$ by $2x$ and we shall get $2x^2$ which will be equal to 10. We shall divide 10 by 2. And then the value of the thing $[x]$ will be the root of 5 which is the first quantity which is a surd and not a number. And for us to find out the second we shall multiply $\mathcal{R}5$ by 2 and this makes $\mathcal{R}20$ and so the second quantity such that $\mathcal{R}20$ is the double of $\mathcal{R}5$, and $\mathcal{R}20$ multiplied by $\mathcal{R}5$ makes $\mathcal{R}100$ as was required of us.

Problem which is solved through the third rule of simple conjunction

Let us look for 2 numbers whose difference is 4 and the lesser growing in the proportion $1\frac{1}{3}$ and the greater in $1\frac{1}{2}$; after these growths let the greater be the double of the lesser, and this will serve for musical instruments. We shall call the lesser $1x$ and the greater will be $1x + 4$, and we shall add to the lesser $1x$ one third since it must grow in the proportion $1\frac{1}{3}$, and the result is $1x\frac{1}{3}$. And we shall add to the greater its half to get $1x\frac{1}{2} + 6$. And since the greater becomes the double of the lesser, because of this $1x\frac{1}{2} + 6$ is the double of $1x\frac{1}{3}$.

And so if we double $1x\frac{1}{3}$ we shall get $2x\frac{2}{3}$ which will necessarily equal $1x\frac{1}{2} + 6$ because *eiusdem sunt dulpicia inter se sunt equalia*. Now we shall proceed with this in the same way. And there will remain $1x\frac{1}{6}$ equal to 6 which is the third rule for the simple conjunction. Let us divide 6 by $1\frac{1}{6}$ and it will become $5\frac{1}{7}$ as the value of the thing which will also be the lesser, and the greater will be this same thing with an additional 4 which is $9\frac{1}{7}$. And the proof has it as follows: because $5\frac{1}{7}$ grows in the proportion $1\frac{1}{3}$ to $6\frac{6}{7}$ and $9\frac{1}{7}$ which grows in the proportion $1\frac{1}{2}$ and will be $13\frac{5}{7}$. Now I state that $13\frac{5}{7}$ is the double of $6\frac{6}{7}$.

By the First Rule of Compound Conjunction

Let us divide 20 by a number where the quotient exceeds the divisor by 4. Let us say that the divisor is $1x$ and therefore the quotient is $1x + 4$. And because multiplying the divisor by the quotient we shall always get the dividend we shall multiply $1x$ by $1x + 4$ and we obtain $1x^2 + 4x$ which will equal 20. And this is the first of the compounds, whose practice is this: half of 4 is 2 which multiplied by itself makes 4 and this 4 added to 20 comes to 24 from which $\mathcal{R}24 - 2$ will be the value of the thing which is the divisor, and because the quotient exceeds the divisor by 4 let us add 4 to $\mathcal{R}24 - 2$ and we shall get $\mathcal{R}24 + 2$. So that 20 divided by $\mathcal{R}24 - 2$ gives $\mathcal{R}24 + 2$ and the proof is that if we multiply $\mathcal{R}24 - 2$ by $\mathcal{R}24 + 2$ we get 20.

By the Second Rule of Compound Conjunction

Let us look for a number which with 20 is equal to its square. We shall let the nuber be $1x$, and adding 20 to it we shall make $1x + 20$ and because the square is $1x^2$ it will therefore be $1x + 20$ equal to $1x^2$. And this is the second rule of compounds. The half of the number of things is $\frac{1}{2}$ which multiplied by itself makes $\frac{1}{4}$. and adding 20 we get $20\frac{1}{4}$ whose root is $4\frac{1}{2}$ and add $\frac{1}{2}$ to this which is the half of the number of things we shall get 5 which will be the value of the thing and this will be the number we are after.

By the third rule of compound conjunction

Let us look for a number which multiplied by 6 makes the same as its square added to 8. Let the number be $1x$ which multiplied by 6 makes $6x$. and its square is $1x^2$ which with 8 makes $1x^2 + 8$ which will be equal to $6x$ and is the third compound conjunction. The half of the number of the thing is equal to 3 which multiplied by itself makes 9, and from this we shall take the number which is 8, and 1 will remain whose root is 1 which we shall subtract from 3, which is the half of the number of the thing and 2 will remain as the value of the thing which is the number we are seeking. Or we shall add 1 to 3 and we will get 4 which could also be the value of the thing

Laus Deo: God be Praised

Chapter Six

Algebra: Contents' Analysis

Where Portuguese mathematics was concerned, the 16th century could justly be described as the century of Pedro Nunes.[52] For a full reappraisal of the development of Nunes' mathematical expertise, an expert in the history of mathematics is needed, with a knowledge of Spanish.[53] However, a few points can be made to set this discovery in its context. First and foremost is the unexpected near-completion of his mathematical inventiveness by the time of his tutorship of the royal Princes. As the two tables in Appendix A reveal, Nunes added very little to his private notes of 1533, where his *álgebra* was concerned, when he converted them to the full-scale Spanish text in 1567. The section on Algebra in the Portuguese text may only contain 24 pages, compared with over 600 in the Antwerp edition, and yet the contents are virtually the same. For his notebook, Nunes, like any competent Mathematics teacher today, could rely on the blackboard or its equivalent, and on plenty of suitable problems stored in his memory; all of which needed to be included in the published version. The printed text starts with a lengthy introduction, in which he sets out the general rules for simple and compound conjunctions. In the Portuguese original this is restricted to two pages, and delayed to a possibly better position, just before the problems.

For the multiplication of integers (A4), reduction of fractions (A6), multiplication of roots (B3), summing of roots (B4), addition of roots (B5) and general rule for dividing by roots (B7), the printed text has added demonstrations, again blackboard material for a class teacher. For proportions and equalization, two pages or so in the manuscript are developed into two full-scale chapters in the Antwerp edition. Finally, he included only five problems in his notebook, compared with the 187 in the printed edition, most of them from his University classes, one would imagine. The Spanish text uses all the same chapter headings and basic formulae, and is full of Lusisms. It seems more than likely that it was a full-scale translation by Nunes of his Portuguese text, based on the skeleton notes that he had jotted down many years before, in his teacher's notebook.

For this publication, Nunes had to enlarge his chapters, with many more examples and demonstrations to elucidate each new lesson, and with a substantial body of self-testing problems at the end. However, the conjugations were perhaps better placed after the roots, together with the rules of algebra, rather than at the start, and his discussions of proportions (equality of ratios) and of equalization, although very brief, are most unexpected forecasts of the *proporciones* and *ygualacion* in the Spanish text. It seems that by 1533 Nunes had already worked out almost his entire programme of algebra, jotting it down in his notebook with just a few equations and problems, for him to enlarge on from memory once he was in front of his class, first as the mathematics tutor to the Princes Henry and Louis in the Court at Évora, and then as Professor of Mathematics at the University of Coimbra.

In his very perceptive study of Nunes' *álgebra*,[54] Joaquim de Carvalho accepted as water-tight the thesis of Dr Gomes Teixeira,[55] who had suggested that Nunes' doctrine of *proporções*, placed in part iii of the second volume of the Spanish text, was based on the *Tratado dos proporções* that formed part of his work *De Crepusculis*, dedicated to King John III in 1541, and published in 1542. However, Joaquim de Carvalho had to admit that the original chronology for the *proporções* could not be worked out for certain, nor fitted into any chronological evolution of his *álgebra*. Even so he thought it likely that all or most of its material was incorporated into his first work on algebra, written, but not published, between 1535 and 1537. He argued that this was suggested by the *Sciencias Mathematicas e*

Cosmografia that were included in his *Tratado da Sphera*, published in 1537. Carvalho also suggested that it was originally a didactic, practical work, with some of Nunes' theory, and his conjectures were surprisingly accurate. His dates, however, are two years late for the first one page discussion of *proporçãos*, which Nunes entered into his teacher's notebook, and then elaborated to nearly sixty folios in his Spanish translation, 34 years later.

It is possible that part or all of his *álgebra* was circulated by Nunes, several years before its publication in Antwerp, among one or two of his French friends who were also mathematicians, like Élie Vinet and Jacques Peletier du Mans (1517-1583). Peletier in fact published one of the letters he sent to Nunes, in the Appendix to his book *In Euclidis Elementa Geometrica demonstrationum libri sex* (Lyons, 1557). In chapter 1 he says: "Here I again mention that Pedro Nunes, a Mathematician of Lisbon, in Portugal, has also translated his algebra into his Spanish language, but I have not seen his book."[56] As this was written in 1554, it suggests that Peletier may have heard of, or even seen, some of the original Portuguese *álgebra*, or more likely of its as yet unpublished translation into Spanish.[57] But it is quite probable that he misunderstood a comment by Nunes expressing his intention to translate the work, or relied on a remark by Élie Vinet who had been shown the manuscript, as was suggested by Joaquim de Carvalho. [58]

Vinet had begun teaching mathematics at the College of Guyenne in January 1541, and served as a lecturer in maths at the Royal College of Arts, in Coimbra, from 1548 to 1555, later becaming Principal of the College in Bordeaux. It seems very likely that he would have become friendly with Pedro Nunes while in Coimbra, where Nunes was the professor of Mathematics throughout his stay, and would have kept up his friendship while teaching mathematics at the College in Bordeaux, from 1556 onwards. For his edition of *Sphaera Ioannis Sacro Bosco emendata* (printed 1552), Vinet translated into Latin Nunes' *Annotação sobre as derradeiras palauras do Capit. dos Climas* that had followed his translation and edition of Johannes de Sacrobosco's *De Sphaera Mundi*, adding them to his own edition.[59] This was the first time that Nunes' writings received any publicity in Northern Europe. In a later work by Élie Vinet entitled: *La maniere de faire les Solaires, que communemant on apele Quadrans* (Bordeaux, 1583), he recommended his readers to study the *De erratis Orontii Finaei* "which my friend Pedro Nunes, the Cosmographer of the King of Portugal, John III, published and had printed at the University of Coimbra in Portugal."[60] This was despite the fact that Finé was a very eminent French mathematician. Vinet was no jingoist.

To modernize the teaching of geometry and mathematics at the tertiary centres like the Royal College in Coimbra, more rigorous explanations of Euclid were needed, and these were provided by Jacques Peletier, with his *L'Algèbre* in 1554[61] and in 1557 his *In Euclidis Elementa Geometrica demonstrationum*. It seems very likely that Nunes, who appears to have shared a kindred spirit with these two French scholars,[62] would have discussed his own discoveries in algebra with them, as well as presenting them with copies of some of his other scientific publications, like his attack on Finé, and his edition of Sacrobosco with its interesting addendum.

Two other major mathematicians of his day were Jerónimo Cardan (1501-1576) and Niccolò Fontana Tartaglia (1499-1557). David Smith argued that in his treatment of roots, with an equation like $x^3 + 3x = 36$ and $x^3 + 9x = 54$, Nunes already showed familiarity with the works of both of these scholars, indicating that Tartaglia's rule was not practicable where one root was easily found by factoring.[63] In his *álgebra*, Nunes criticized Tartaglia's rule for resolving the equation $x^3 + ax = b$, expressed by the Italian in a nine verse poem.[64] However, Cardan's *De Subtilitate* was published in 1550, and Tartaglia's *Tratado di numeri e misure* in 1556-60, both of them long after Nunes' original *álgebra* had been included in his teacher's notebook. Like all of the mathematicians of his day, however, Nunes failed to appreciate the value of negative quantities in his algebra, despite the suspicions of Luc Paciuolo, a basic concept that in fact was only fully accepted by mathematicians in the 17th century. To Nunes, one-time professor of Logic, the negative quantities were considered an absurdity, an absolute contradiction.[65]

Nunes was also somewhat out-of-date in his algebraic notations. He ignored those used by the German algebraists, the x, x^2, x^3 etc., and showed no sign of having read Michael Stifel's *Arithmetica Integra* (Nuremburg, 1544).[66] Instead, he followed the Italian notations, with Co.2. Ce.4. Cu.8. etc., as in Paciuolo's *Sûma de Arithmetica, Geometria* ...(Venice, 1494, and 1523[2]), and in Jerónimo Cardan's *Practica Arithmeticae* (Milan, 1539). For such an Hellenist as Nunes, it is surprising that he did not even follow the lead of the early mathematicians, Euclid and Archimedes, who used numbers and lines, with letters only to distinguish the lines. With lettering only, there was extra complexity, which could lead to some difficulties, especially if letters were poorly written, and it was more prone to ambiguities. Fortunately Nunes wrote with the clarity of a humanist, and the conversion of his lettering to modern numerical notation was not too hard. However, it may have caused some problems for his young students, and for readers used to the German system of notation.

And yet where the distinction between geometry and algebra was concerned, Nunes was well ahead of his contemporaries. The French mathematician Jordan de Némore, in his *Arithmetica decem libris demonstrata* (Paris, 1514), had rejected Euclid's use of lines and letters, and concentrated on the letters alone. His work was the first to do so systematically. But his novel approach went virtually unheeded for nearly 200 years, until the work of Pedro Nunes.[67] This was one of his really great achievements, as Bosmans demonstrated with his algebraic resolution of $c\sqrt{b} = a$, the reduction of roots to the same power or form, which ends with Nunes' valid claim that his

As we have seen, for his research Nunes was ready to translate many original texts written in several different languages, such as Ancient Greek, Latin, Italian and Arabic. For example, for his *Tratado da Sphera* (1537), he translated book 1 of the Γεωγραφικὴ Ὑφήγησις by Ptolemy of Alexandria (*fl.* 121-51 AD), Sacrobosco's *De Sphaera Mundi,* to which he added marginal notes to explain and correct the text,[73] and Purbach's astronomical treatise.[74] In the dedication of his *De Crepusculis liber unus* to King John III (1542), Nunes shows that he had started on a translation and edition of another very interesting scientific work, Vitruvius Pollio's *De Architectura* , a treatise in 10 books on architecture and engineering, mostly based on Greek originals, and one of the few such works to have survived from Antiquity. Nunes pleads ill-health, plus daily classes with Prince Louis explaining the works of Aristotle, as an excuse for having delayed his edition for so long, only half of it having been completed; but this sounds a bit like special pleading.[75] Vitruvius covered the rules of proportion in books 3 and 4, and geometry and astronomy in book 9, material which would have been of use to Nunes, but he may have felt that there was no real need for the edition to be completed. However, it is likely that he was encouraged and helped to undertake this work on Vitruvius, especially book 8 on water-supplies and aqueducts, by his friend and fellow royal tutor, Resende, who presented the King with two treatises on Aqueducts. Unfortunately neither they nor Nunes' work on Vitruvius, if ever completed, have survived today. Perhaps one of them may turn up in another miscellany gathering dust in the archives of the Municipal Library of Évora.

Nunes was certainly a prolific writer in his early years, and it is not hard to see why he felt no need to publish the materials in his teacher's manual, when one considers how many of his other works never appeared in print, sometimes because they had been up-staged by the time that they were finished. Besides the Vitruvius, there are five such treatises listed by Guimarães,[76]and one other, as follows:

(1) *Tratado de Geometria dos triangulos spheraes* (Treatise on the Geometry of spherical triangles) Nunes mentions, at the end of his *Tratado da Sphera,* that he had written most of this treatise before he was sent books on the same subject by Gebre and Monteregio from Germany. After reading them, however, he did not destroy what he had written.

(2) *Tratado sobre o astrolabio* (Treatise on the Astrolabe). There may be a copy of this (in a late XVI century hand) in the Municipal Library of Porto (Ms no. 250), with 100 chapters, on 150 sheets, according to Guimarães.

(3) *Tratado do planispherio geometrico* (Treatise on a Geometric planisphere). Probably based on Euclid.

(4) *Tratado da proporção ao livro V de Euclides* (Treatise on Proportion in Euclid *Elements* Book 5). This may have been reduced and included in his attack on Finé.

(5) *Tratado da maneira de delinear o globo para uso da arte de navegar* (Treatise on the way of drawing the globe for use in the art of navigation).

(6) *Defensão do Tratado da rumação do globo para a arte de navegar* ((Defence of his treatise on the plotting of the globe for the art of navigation).[77]

As we saw in the introduction to this book, in addition to his pioneering work on magnetic variation and on the divergence of the meridians, Nunes was responsible for four important nautical inventions;. He was also the first to solve the extremely important problem of calculating one's latitude at any time of day, in his *Tratado em defensam da carta de marear.* [78] Despite the fact that he did not know tangents, and made no use of sines, Nunes was responsible for many extraordinary inventions, mainly due to his early brilliance in understanding and applying the rules of algebra. Using its formulae, he made a major contribution to nautical science, while giving his country's mariners the necessary charts and nautical equipment for them to cross such vast oceans and discover and map so many distant countries. Without his life-long service to his country, it seems likely that many more of the wide-ranging explorations of the Portuguese would have been usurped by Spanish galleons.

Chapter Seven

Poems: Contents' Analysis

The recently discovered collection of Greek and Latin poems written by Pedro Nunes was published in 1991, very soon after I had transcribed them.[79]   Of the sixty poems, 36 (222 verses) are in Ancient Greek, and 24 (204 verses) are in Latin.   About half of the poems are on the death of Nunes' patron, Prince Louis, written long after Nunes had left Évora for his chair of Mathematics in Coimbra. Including these, as many as 39 are on the Royal family, all of them eulogistic in nature, as one would expect from a Royal tutor.   Most of the rest are on religious topics, again normal at a time when the Church played such a major part in men's lives, especially in those of the pious children of King Manuel and of his successor, King John III.   Some are on contemporary pedagogical matters, and one is on the City of Lisbon.

Far the most interesting poems for Nunes' early years in the Royal Court are two in Latin that link him with a fellow Royal tutor, António Pinheiro, the future Bishop of Miranda and Leiria. In old age he became the spokesman for Portugal's Council of State (in 1581), which readily supported Philip II's right to the Portuguese throne, left vacant at the death of Sebastian in the sands of Morocco.[80] These two poems also help to explain the short but interesting Latin eulogy that follows, dedicated by the same António Pinheiro to the 1541 original of Nunes' *D e Crepusculis*.[81]   He was born in Porto de Mós (hence his Latin name Pinus Portodemoeus) probably in the year 1511, given his initial enrolment at Ste Barbe in 1527 as a *puer* (16 maximum).[82]   The son of Pedro Vaz de Couto and Leonor Alvares Pinheiro, he served as an apprentice to the Bishops Diogo Pinheiro and Gonçalo Pinheiro, before being sent to Paris in 1527 as a Royal scholar. There, after

after a moral and literary education from Jacques Strébée, and three years of philosophy under old Diogo de Gouveia, Principal of Ste Barbe, he became the first such scholar to publish a learned work, editing Quintilian's *Institutiones Oratoriae* Book III (Paris, 1538)[83]. In the Latin dedication, he sings the praises of Gouveia and of Quintilian, two major influences on the rest of his life. Like Gouveia, however, he might be seen as a traitor to his country, the former betraying to the Inquisition three leading humanists at the Royal College, George Buchanan, João da Costa and Diogo de Teive, and thereby undermining the country's humanistic revival, the latter betraying his country to Philip II of Spain. Like Quintilian, Pinheiro lectured on rhetoric and edited and translated key works on public speaking. For example, his edition of *Inst. Orat.* III and his translations into Portuguese of the younger Pliny's *Panegyricus* (for the King's benefit), his letter on the speech (III.18) and Cicero's long letter to his brother Quintus (I.1) He also became his country's most eloquent public speaker and preacher, and a trusted adviser first to King John III and later to the young King Sebastian.

After obtaining his doctorate, Pinheiro was lecturing in Ste Barbe on rhetoric, but was chosen by Frei André as tutor for the young noblemen at the Royal Court, and, had to travel down to Évora, at the King's order, possibly in the company of Frei André.[84] Luís de Matos suggests 1542, but 1540 seems more likely.[85] Soon afterwards he replaced Damião de Gois as tutor to Prince John, who was sworn in as heir to the Portuguese throne in Almeirim on 30 March 1554, with an official prayer by Pinheiro.

On 16 July 1550, Pinheiro was appointed chief chronicler to the King, but excused himself when asked to write the chronicles of Kings John II and Manuel I, leaving the task to a far better historian, Damião de Gois, who was appointed on the recommendation of Cardinal Henry. He was chosen to give the official prayers at Sebastian's acclamation, and on the occasion of the translation of the bones of Manuel I and Maria. His relationship with Jerónimo Cardoso can be seen in the letter. In a speech in 1563, after the King's death, he justified the move of the University of Lisbon to Coimbra, which many academics had opposed, using arguments first put forward by Ludovico Vives in 1531. In the dedication of his work *On Education* to the King, Vives had argued that Coimbra was healthier, and clear of commerce and sea-trade.[86] His circle of humanistic friends also included Rodrigo Sanches and Joanna Vaz, and in 1547 he was praised by the Latin poet Diogo Pires.

By 1564, Pinheiro was a very active Bishop of Miranda, founding a charity at Algoso and a Jesuit college at Bragança, although he later supported Queen Catharine in her attempt to combat the growing influence of the Jesuits.[87] In 1574 he accompanied Sebastian as chaplain during his first foray into North Africa, but attacked it as risky and pointless. He was sacked by the young King, but still spoke out against his second and fatal attack on Morocco. After Sebastian's death, he was one of the judges who rejected the claim of Don António, the Prior of Crato, to the throne of Portugal, and was then moved to the Bishopric of Leiria, where he stayed during the turbulence of the brief civil war. A firm supporter of the ideal, God-appointed King, he fully endorsed Philip II's claim to the throne, praying at his acclamation in April 1581 and at the swearing in of Prince Diogo, but died soon afterwards, on 9 December 1582 (or 9 January 1583), in his early seventies, probably in Lisbon. His poem to his fellow-tutor, Pedro Nunes, can be seen below. Two other Latin poems by Pinheiro appear in Appendix D. A man of great ability and great influence, Pinheiro deserves a full-scale biography.

Carmen   in   laudem   operis

Cynthia quae rapidis nocturna crepuscula bigis
   proferat, aut rutilos sol ubi pingat[88]equos,
quam certis medius constet regionibus aër,
   aethereo quae sint sidera fixa polo,
omnia sollerti vestigans ordine Petrus   5
   Nonnius, Herculea dat tibi, lector, ope.

Tolle humiles animos terrarumque exue curis
pectora; non magnus magna libellus habet.

Poem in praise of the Work

What night shadows are spread out by the Moon on her
   swift chariot? Where does the sun paint his horses red?
Of what fixed regions does the middle atmosphere consist?
   What stars are anchored in the heavenly sky?
All these questions Nunes investigates with wise logic,   5
   giving you answers, reader, with Herculean power.
Raise your humble minds, lift up your hearts, clear of the
   Earth's cares. A small booklet, it has great contents.

This poem was written about eight years after a period in
which it appears that he shared the teaching of the royal
Princes with Pedro Nunes.   From the newly discovered
poems, written when both were tutoring in Évora, or soon
afterwards, some very interesting facts emerge, not
entirely to the credit of the future Bishop.   The first poem
appears to be ironical, as it dramatically describes an
auction of the Muses, due to their hunger, that can only be
cured by the King's patronage.[89] It reads as follows:

Auctio Musarum, prae fame

Auctio fit.   Loculis propera, mercator, apertis.
   Quando? Ubi? Palladia sole cadente domo.
Quae merces? Musae. Quis clamat?   Praeco canorus.
   Quis studet ingenuas vendere?   Serva fames.
Sol cadit. Ecce patent aedes, stat turba sororum,   5
   statque frequens emptor, praeco canore tono.
Quis doctas emit Aonidas ter mille talentis?
   Tanti emerem saturas?   Non emo.   Pinus emet.
Auctio ne fiat, decies ter mille talenta
   Lysiadum dederit Pieridumque pater.   10
Hinc procul ad merces solitas, mercator, abire
   festina; tu etiam, sordide praeco, tace.
Et tu, causa mali, diro sub sidere Cancri,

Aethiopas vendes, invidiosa fames.
Aonidas spernant quorum sunt facta silenda;   15
   regis sunt Musis gesta canenda novem,
qui primus natas fecit natasque vocavit,
   atque decus sceptri credidit esse sui.
Serva fames fugito. Regis tu vendere tanti
   ipsius in regno pignora cara velis?   20

Auction of the Muses, due to their hunger

An auction is taking place. Hurry, merchant, with purse
   open. When? Where? At the Muses' house, at sunset.
What's for sale? Muses. The crier? A raucous auctioneer.
   Who is keen to sell free-born girls? Servile hunger.
It's sunset. See, the house is open, there stands a group of 5
   sisters, lots of buyers and a loud-mouthed auctioneer.
Who buys learned Muses - for 3,000 talents? Would I buy
   satires for so much? I'm not buying. Pinheiro will buy.
To stop the auction, the father of the Portuguese race and
   of literature will give ten times 3,000 talents.   10
Merchant, hurry, go far from here, for your normal
   goods, and be silent also, sordid auctioneer. And you,
invidious hunger, the cause of evil, shall sell off girls
   from Ethiopia, beneath the grim star of Cancer.
Let those whose deeds deserve silence spurn the Muses.   15
   The King's deeds must be sung of by the Muses nine,
for he first created and summoned the young girls,
   and believed them to be the glory of his crown.
Let servile hunger flee. Would you want to sell the dear
   children of so great a King in his very kingdom?   20

When he jokingly asks Pinheiro to make a higher bid for
the Muses, it may suggest that his family was wealthy, and
as he had relatives who were Bishops, and the funds for his
own future Bishoprics, it may well have been so. However,
Pinheiro had earlier gone to study in Paris as one of King
John III's royal scholars, before returning to tutor first
the noblemen, and then young Prince John, at Portugal's
Royal Court. This does not suggest a family of high nobility

or of great wealth. Rather, Nunes seems to be teasing him, perhaps for putting on airs. He must have been a capable Classical scholar, however, with a special interest in the theory of education, to judge from his commentary on Book II of Quintilian's *Institutiones Oratoriae*, published in 1538. He also exchanged letters with leading Portuguese humanists of the day, like Jerónimo Cardoso (see Appendix D) and Rodrigo Sanches. But a chaplainship to King John III seems to have been the catalyst in his rise to a position of great power and influence in his country's future.

King John III's support for the Arts was first shown by his generous scholarships that enabled promising young Portuguese students (like António Pinheiro) to study at the College of Sainte-Barbe in the University of Paris. The college was under the control of old Diogo de Gouveia (its head from 1520 to 1548), cantankerous but much admired by many of his pupils, especially young António Pinheiro. The mostly Portuguese students lived Spartan lives, but heard lectures from some of Europe's finest teachers, including old Gouveia, Jacques Louis Strébée, André de Gouveia (future Principal of the famous Collège de Guyenne) and George Buchanan, tutor to Mary Queen of Scots. As Luís de Matos has well shown,[90] its graduates rewarded the King's interest and investment in the College, providing his country with well-trained lawyers, academics, doctors, priests, teachers and governors for Portugal and its rapidly expanding overseas empire.

Far more significant, however, was the King's rich endowment of his Royal College in Coimbra, opened on 21 February 1548, where tempting salaries and royal pressure encouraged most of the staff (and many students) at the Collège de Guyenne to transfer from Bordeaux to Coimbra, with André de Gouveia as its Principal.[91] By April 1548, there were 1,000 students at the Royal College, including the brilliant Portuguese playwright and poet, António Ferreira. The highly paid staff included George and Patrick Buchanan, Diogo de Teive, João da Costa, the brilliant Aristotelian scholar, Nicolas de Grouchy, the

Latin playwright, Guillaume Guerente, the textual critic and mathematician, Élie Vinet, and the Greek scholar, Arnaldo Fabricio. Unfortunately an embittered old Gouveia had thereby lost his supply of Portuguese students, and then lost control of his beloved College of Sainte-Barbe, and he had quarrelled bitterly with his nephew André. To get his revenge, he noted down many trivial cases of heretical tendencies shown by the new College's leading academics, André de Gouveia, George Buchanan, Diogo de Teive and João da Costa, all of them full of Erasmian ideals, and passed them on to the Inquisition. André was a man of very high standing, and a personal friend of the King, but he died suddenly and most unexpectedly from a fever. Evidence from the College pharmacist during a later enquiry suggests that the real cause of his sudden death was arsenic.[92] The other three humanists were arrested at once, and after many worrying months in solitary confinement, they were finally tried and imprisoned. The Royal College became a College of Arts, and a chastened Teive its last Principal, before it finally came under Jesuit control, in August 1555.

The sociological references in the poem are also of interest. The procedure for an auction, for example, is realistic, held at sunset, after a hot day, and run by a loud-voiced auctioneer, although his trade is much despised by Nunes (perhaps following Juvenal again). The nine young ladies are saved by the King, but African (Moroccan) young girls are readily suggested as an alternative. This points to the Portuguese slave-trade which encouraged some of man's most barbaric treatment of fellow man, especially when African youths were kidnapped, chained and packed into ships like sardines for the American market. In Portugal, both the Church and the humanists, with a few exceptions, condoned this lucrative but inhumane trade in slaves, and readily accepted the degrading system of slavery in their own society. The way in which it could corrupt even a pious man of God can be seen in the life of André de Resende, who had several slaves, and in his sixties had sexual relations with his housekeeper and, it seems, with a female mulatto slave. The first gave birth to a son, Barnabé (1559-1596), whom he placed in service under the Duke of Aveiro, and enriched with most of his goods and chattels. The mulatto offspring of the second, called Maximo, proved too rebellious, and too close to Resende for comfort, who in his Will ensured that Maximo would be sold well away from Évora, in case he was a threat to Barnabé's inheritance.[93]

The second poem is dedicated to Pinheiro, and reads as follows:

Ad Antonium Pinum, virum ornatissimum

Pine, decus patriae, cultor facunde Minervae,
*   ingens Pierii gloria, Pine, chori.*
Ad iuga Parnassi iuvenes, nisi praemia dentur,
*   incedunt lento difficilique gradu.*
Rarus qui manibus pedibusque evadat ad alta   5
*   culmina, quae placido veris honore nitent.*
At multi vel fraude loci vestigia retro
*   vertunt, vel pavido sollicitante metu.*
Praemia si dentur, currunt, si praemia regis
*   nomine Ioannis nobilitata, volant.*   10
Tu cupis, egregio patriae succensus amore,
*   eius ut actutum pignora cara volent.*
Et cupis expertus, nam quondam praepete penna
*   vectus es excelsi montis ad alta iuga.*
Nos etiam cupimus, quibus est tam dulce volantes   15
*   cernere discipulos, quam stimulare grave.*
Sed tamen implumes largiri haud possumus alas;
*   texere quas verbis nitimur, aura rapit.*
Ergo tuo monitu Parnassi, rex, decus ingens,
*   discipulis nostris Daedalus esse potest.*   20

To António Pinheiro, a most distinguished man

Pinheiro, glory of your homeland, eloquent lover of
    learning, great glory, Pinheiro, of the Pierian chorus.
Young men approach the hills of Parnassus with slow

and difficult steps, unless they are given rewards.
Rare is he who emerges on his hands and feet at the   5
    highest peaks, gleaming with Spring's placid beauty.
But many turn their footsteps backwards, either due to
    the place's deceits, or worried by numbing fear.
If rewards are given, they run up there, if rewards made
    noble by the name of King John, they fly up there.   1 0
Inflamed by your splendid love for your country, you want
    those dearest to him to fly up there at once.
And your desire is based on experience, for you were once
    born on rapid wings up to the mountain's high ridges.
I desire it too; I find it just as delightful to see pupils of   1 5
    mine flying up there, but find it hard to stimulate them.
And yet I cannot bestow featherless wings on them;
    those I try to weave with words, the wind blows away.
Thus, with your advice, our King, great glory of Parnassus,
    can become another Daedalus for our pupils.   2 0

As father of Icarus, the prototype craftsman, Daedalus,
wove wings for himself and his son to use in escaping
from King Minos' labyrinthine palace on Crete.
Unfortunately his son flew too close to the son, and his
waxen wings melted, plunging him in the Icarian sea. No
such premature death was implied by Nunes, although
Resende's favourite pupil and patron, Prince Edward, was
to die prematurely, in his mid twenties, as had his elder
brother, Cardinal Alphonse, six months before, aged only
32. Again Nunes sings the praises of the King, who
attracted outstanding teachers to his Court at Évora, and
rewarded them very well, to judge from Nicolas Clenardo's
reception,[94] Resende's Will[95] and the many generous
endowments received by Nunes.[96]

The poem starts by flattering Pinheiro, who had climbed
Parnassus with ease and had mastered the Classical
languages and literature, achieving great glory. However,
this academic success had encouraged Pinheiro to be over-
ambitious for the aristocratic and royal pupils entrusted to
his care, all too ready to let them try out their wings as
scholars, before they are fully trained; unlike Pedro, who

finds it hard to stimulate them, but provides no easy path to success, and no awards for featherless wings. This may refer to the extra difficulty of Nunes' scientific subjects, cosmography, geometry and algebra, and perhaps to his loss of a first-class mathematician (after 10 years of tuition), Prince Henry, who chose to concentrate on his theological studies under Clenardo, as a young Archbishop of Braga, later to become the Archbishop of Évora and Inquisitor General.[97] To his credit, however, he chose André de Resende as his chaplain and theological adviser, and donated his palace in Évora to the newly formed Jesuits, as a home for their very first University.

The two poems that mention Pinheiro are the longest of his Latin poems, and were placed at the start and end of the 24 poems, amounting to 204 verses. There are also 36 poems in Greek, amounting to 222 verses, an extraordinary and almost unique achievement at this time, as will be shown below. The remaining 22 Latin poems are on the death of Prince Louis, far the saddest loss in Nunes' life, it seems, as are nine of the Greek ones. These can be seen in my articles on his poems in *Euphrosyne*, but some are of special interest for the life and character of Nunes' patron, Prince Louis.[98] The first is on his rejection of worldly goods:

Epitaphium   Principis   Lodoici

Crediderat Christo gemmas Lodoicus et aurum,
*   auratos etiam purpureosque toros.*
Pocula crediderat precioso facta metallo,
*   sed quibus ex arte gratia maior erat.*
Crediderat phaleras saturatas murice et auro,   5
*   et Libyca pictas artificique manu,*
regificas etiam vestes pulchrosque tapetes,
*   certantes ventis alipedesque suos.*
Crediderat ludos, circum, spectacula, mensa,
*   crediderat vitae gaudia cuncta suae.*   10
Cum nil restaret fere iam quod credere posses,
*   o decus, o generis gloria magna tui,*

astra petis, dives sola vel sorte futurus.
Accedant sorti fenora: Croese, tace.

Epitaph for Prince Louis

Louis had entrusted his jewels and gold to Christ,
    and his golden and purple couches also.
He had entrusted his goblets, made from precious metal,
    but given greater beauty through artistry.
He had entrusted necklets imbued with purple and gold,   5
    and painted by a Libyan artist's hand,
and regal robes as well, and beautiful tapestries,
    and his own horses, that were rivals to the winds.
He had entrusted sport and circus, spectacles and banquets,
    he had entrusted all the pleasures of his life.   10
When almost nothing was now left for you to entrust,
    Louis, the splendour and great glory of your family,
you sought the stars, to be rich hereafter in luck alone;
    let interest accrue to your luck. Croesus, keep quiet.

The symbols of regal display are well depicted in lines 1-8, and the final conceit gives added point to the poem. Croesus, King of Lydia (560-546 BC), was proverbial for his wealth, but it pales into insignificance beside the heavenly riches of Louis, which earn so high a rate of interest. In fact Louis was the most pious of all the Princes, and came close to joining the Jesuits. According to Jesuit sources,[99] he gave up all of his rich possessions and devoted his life to Christian doctrines, making a vow to God of chastity, poverty and obedience. After persuading his brother, Cardinal Henry, to donate his palace in Évora to the Order, Louis visited the sick in their new college, and even held men's arteries. His funeral echoed his life, as the eminent Jesuit, Pedro Perpinhão, sang his praises in a polished Latin oration in Coimbra's College of the Arts, in December 1555.

Nunes' Greek and Latin poems delivered much the same message. In them, Louis appears as a true Christian, comparable with St Francis and St. Jerome, as he rejects earthly pleasures and possessions for a life of prayer and regular communion. As a young man he had been forced to accompany Charles V during his assault on Tunis. During the fighting, the Emperor was highly impressed by the Prince's courage and leadership qualities, but to Louis it was primarily a holy mission. An aristocratic courtier and religious Latin poet of that time, Jorge Coelho, dedicated a twelve verse poem to the *De Sphera* of Pedro Nunes, and they both wrote poems about the sack of Tunis. Among Coelho's works published in 1540, under the heading of his masterpiece, *De Patientia Christiana*, there is one of 91 hexameters dedicated to Prince Louis, in which he contrasts the undisciplined troops from Germany, Spain and Italy, who piled up looted treasures while drinking and murdering, to the Prince who only sought the glory of God:

> *Tu pius interea et sedato pectore felix*
> *caelum animo, Lodovice, capis.*

> "Meanwhile you, Louis, holy and fortunate,
> with a peaceful heart, think only of Heaven".

His great act of holiness was to bring back neither gold nor treasures, but a statue of the Blessed Virgin Mary, the Mother of God, that he had discovered in Tunis.

The same antithesis appears in another of Nunes' Latin poems:

De Principe Lodovico

> *Victrici quantum dextra polleret et armis*
>   *quam vehemens, acer terribilisque foret,*
> *hostes ut magna Lodovicus vinceret arte,*
>   *bellum si quando posceret ingenium,*
> *Caecareae norunt aquilae et Tunisia regna,*   5
>   *ipse etiam noras, Aeneobarbe ferox.*
> *Infernas sed enim magis est vicisse phalanges;*
>   *vicisset si orbem, fecerat ille minus.*

On   Prince   Louis

How strong he was, with his victorious right arm, how
   passionate, keen and terrifying with his weapons,
how Louis conquered the enemy with great artifice,
   whenever the war demanded intelligence, is known
by the eagles of Caesar and the kingdom of Tunis, and   5
   and you yourself know it too, fierce Barbarossa.
But it is better to have overcome Hell's legions; even if he
   had conquered the world, he would have achieved less.

The other poem that mentions Tunis and his military
prowess serves also as a thumb-nail sketch of the Prince.
It too consists of seven elegiac couplets, and reads as
follows:

De   obitu   Principis   Lodoici

Ille suis carus, tam formidatus ab Afro,
   in quo certabat cum probitate decor,
ille animis opibusque potens, tot regibus ortus,
   excellens armis, artibus, ingenio,
ille sui iudex et servantissimus aequi,   *5*
   floruit in cuius pectore sancta fides,
ille habitans caelo, cuius dulcissima cura,
   cuius nobilitas maxima Christus erat,
ille, teres mundi instar habens morumque magister,
   qui tenebris hominum lampadis instar erat.   *1 0*
Ille suae totum qui afflabat floribus orbem
   virtutis, terris cognitus atque polo,
occidit, atque suo liquit nos funere maestos,
   occidit;   erravi, vivere coepit enim.

On   the   Death   of   Prince   Louis

That man, dear to his own people, so feared by Africans,
   in whom beauty competed with honesty,
that man, powerful in spirit and wealth, born from so
   many Kings, outstanding in arms, in the Arts and in
talent, that man, critical of himself and most observant   5

of justice, in whose heart holy faith has flourished,
that man, dwelling in Heaven, for whom Christ was
   the sweetest love and the greatest nobility,
that man, grasping the world's round design, a teacher of
   morality, who was like a lantern for men's shadows,   10
that man, who covered the whole world with the flowers
   of his virtue, well known on Earth and in Heaven,
he has died, and has left us in mourning over his death,
   he has died. I have erred, for he has begun to live.

The use of *instar* in v. 9 as a noun ('design') may suggest
Nunes' lessons on cosmology (and shadows), but in its
second appearance in the pentameter, it is an adverb
taking the genitive, common in poetry and meaning "just
like". Again the final couplet is pointed, with the
repetition and correction of *occidit*. As has been shown,
Nunes used great dexterity and considerable artistry in
composing his Latin poems. Not surprisingly, he seems to
have found Greek elegiacs far harder to write, often
repeating phrases he has coined, and at times struggling
with the metre and different Greek dialects.

Many of the other Latin poems are of interest, and there is
an historical one that deserves to be included here, in
which Prince Louis is depicted entering Heaven where he
is met by close relatives who had lately died, especially his
father and his three brothers, all but one born after him.

De obitu Principis Lodoici

Maximus Emanuel super et Garamantas et Indos
   qui Lusitanum protulit imperium,
occurrit nato venienti ad limina caeli,
   eius et amplexu gaudia mira capit.
Occurrunt etiam fratres dulcesque sorores,   5
   ingressum ut celebrent, o Lodoice, tuum.
Hic rutilans Fernandus ait: "Dulcissime frater,
   haec sunt virtuti debita sceptra tuae."
Mox fulgens radiis Eduardus luminis aurei:
   "Venisti tandem, mi Lodoice," refert.   10

Dehinc sacer Alphonsus Phoebo fulgentior inquit:
O propera, o felix, o Lodoice, veni."
Mox neptes patruo occurrunt carique nepotes,
quos ornat pulcher sidereusque decor.
Sceptrifer hos inter iuvenes novus incola caeli: 1 5
"Serus, sed felix, mi Lodoice, venis.
Quam miseret patris cari patruique verendi!
Nunc melius cernes carcere quo iaceant."
Post haec Emanuel rex, cinctus prole beata,
Suspiciens Christi numina clara Dei: 2 0
"Quas" inquit "possim Pater o tibi dicere laudes,
in quem tam largas, optime, fundis opes?
Quos mortis genui servos in carcere terrae,
in caelo reges tu mihi restituis."

On the Death of Prince Louis

Most noble Manuel, who extended the Portuguese empire
   over both the Africans and the Indians,
met his son, as he approached the threshold of Heaven,
   and enjoyed wonderful pleasure as they embraced.
His brothers and sweet sisters came to meet him also,   5
   so as to celebrate your entry, dear Louis.
Here gleaming Ferdinand spoke: "Sweetest of brothers,
   these sceptres are indebted to your goodness."
Soon Edward, gleaming with rays of golden light, spoke
   to him: "At last you have come, my dear Louis."   1 0
After him, holy Alphonse, more dazzling than Apollo,
   cried: "Hurry here, fortunate Louis, come here, come."
Soon nieces and dear nephews ran up to their uncle;
   a beautiful, starry brightness embellished them.
Amid these youths a kingly one, newly come to Heaven   1 5
   cried: "You come here late but fortunate, my dear Louis.
How we should pity my dear father and venerable uncle!
   Now you see more clearly in what a prison they lie."
After this, King Manuel, surrounded by his blessed
   offspring, looked up to the bright divinity of Christ   2 0
our God and said: "What praises could I give thee, almighty
   Father, who pourest such lavish riches upon me.
Those whom I begat as slaves to death in Earth's prison

are restored to me by you as Kings in Heaven."

The pointed finale is very much part of Nunes' literary style, similar to the last couplet of a sonnet. Of the many Princes and Princesses waiting for Louis in Heaven, when he had died on 27 November 1555, only one, his father King Manuel, had been born before him. Ferdinand, born in 1507, died in 1534, 21 years before his brother Louis. Edward, born in 1515, died in 1540, a few months after Alphonse, who had been born in 1509. The kingly youth was the Prince of Denmark, son of Christian II and Isabella, the sister of Charles V, and foster-son of the Emperor. He died in 1532, while he was travelling through Germany under the personal care of Charles V, who was on a mission with his mainly German army to save Vienna from Saladin's besieging forces. The bitter loss was well expressed in an *Epicedion* written by André de Resende, in honour of the youth. His father died in a prison in Kallundborg in 1559, four years after Louis' death, after being in prison from 1520, when his cruel tyranny and massacres had led to a popular uprising in Sweden. His two "sweet sisters" were Isabel (1503-1539), and Beatrice (1504-1538). Notably absent was the successor to Manuel, King John III, born in 1502, who lived until 11 June 1557, outliving all nine of his children, and John III's younger brother, the Inquisitor General, Cardinal Henry, who was born in 1512 and survived all of the many Princes and Princesses of three generations. He finally become King himself, albeit briefly, from 1578 to 1580, when he too died, leaving the throne for the Spanish king Phillip II to usurp, with his blessing.

Nautical imagery is to be found in several of Nunes' poems, and this is not at all surprising when one considers his close links with mariners and sea charts, even though it seems that he never went to sea himself. For example, the second poem in the collection, on the death of his patron Prince Louis:

De obitu Principis Lodoici

Cum nihil in terris princeps Lodoicus haberet
    mortale et vitam crederet exilium,
solus et in sacro regnaret pectore Christi
    dulcis, mellifluus, religiosus amor.
in celsum cupidus tendebat carbasa caelum,   5
    attentis oculis sidera clara notans.
Ipse gubernaclo affixus perstitit et haeret
    inceptum cursum nocte dieque sequens.
Iactatum pelago Christus miseratus amicum
    immisso Zephyro sistit in arce poli.   10

On the death of Prince Louis

Since Prince Louis possessed nothing on Earth that was
    mortal, and believed that life was an exile,
and when a charming, honey-sweet, religious love of
    Christ held sway alone in his holy heart,
he set his sails eagerly, towards lofty Heaven,   5
    marking the bright stars with attentive eyes.
He persisted on his own fixed to the tiller, and holding fast
    followed the course undertaken by night and day.
Christ felt pity for his friend, tossed about by the ocean,
    and sent a wind to halt him at the citadel of Heaven.   10

Fixed and clinging to the tiller, like Odysseus as he sailed
past the Sirens, Prince Louis is depicted as a man sailing on
his own, with only the stars to guide him, up to Heaven,
again like Odysseus in his final quest for Ithaca. The
Prince's ready acceptance of death is stressed here and in
many of the other poems, as is that of Prince Edward in
Resende's prose account of his death scene. It seems that
all of Manuel's sons were extremely pious, although strong
supporters of Erasmus and his ideals, whose writings were
also the favourite reading material of King John III and
his wife, Catharine. The King also opposed the entry of the
Spanish Inquisition into his country, but in his later years
he had to succumb and witness its barbaric autos-da-fé,
staged by his brother, Cardinal Henry. Nor was he able to

prevent the loss of the Royal College, founded by him with pride and enthusiasm in Coimbra, to the Company of Jesus.

Unlike most of Portugal's New Christians, who had to leave the country or risk being burnt at the stake, Pedro Nunes seems to have maintained a close friendship with Prince Henry, and André de Resende likewise. The latter first had to renounce his love for Erasmus, and then serve for many years as chaplain and religious adviser to the Cardinal, marking his orthodoxy with several poems eulogizing the faith and generosity of his patron. As we have seen, Nunes dedicated his great work on algebra to the Cardinal, and composed two of these poems in his honour, one in Latin and one in Greek. The first shows Cardinal Henry's genuine grief at the death of his dearest brother, Louis :

Ad Enriquum Principem, de obitu Lod. Principis

Enriquus cari maneret dum funera fratris,
  dilexit vita quem magis ipse sua,
fluctuat ancipiti cura ter maximus heros,
  heros purpurei lausque decusque chori,
anne magis lucem celebret laetissimus almam   5
  ortus quae fratri praestitit aethereos.

To Prince Henry, on the death of Prince Louis

While Henry was weeping over the death of his dear
  brother, loved by him more than his own life,
the almighty hero, the hero and honour and glory of the
  purple choir, wavered with a double concern,
whether he should not rather celebrate most joyfully the 5
  blessed day that gave a Heavenly birth to his brother.

The second poem is a eulogy of Cardinal Henry, the first really active Inquisitor General in Portugal, a cheerful poem written many years before, when Louis was still alive. The English runs as follows:

To   Prince   Henry,   Cardinal

The sceptre-bearing King is brother to Henry, and is
    a beacon of justice, of wisdom and of piety.
But Henry himself is the most egregious archbishop
    of Christ, equal to his brother in intelligence.
Hail, people of Portugal, rejoice, greatly blessed race,   5
    to whom God has given two such fine shepherds.

Only one of his Greek poems was dedicated to a living
Prince Louis, written during his early days as royal tutor
in Évora.   It links the transience of empires with the
Prince's preference for dust-free riches of the soul.   The
cliché is given a neat personal touch:

To   Prince   Louis

Even the beautiful walls of mighty Babylon collapsed, and
    the Mausoleum memorial and work of the Pyramids.
And you, knowing this, Louis, and knowing precisely with
    what sun the land of Portugal is illuminated,
are seeking only the riches of the soul, correctly I am   5
    sure, for dust has never covered them over.

The appearance of any original Greek poems in mid
sixteenth century Europe was most unusual, and for a
fluent Latin poet to compose a third more poems in Greek
than in Latin is even stranger.   Most were written while
he was still tutoring in Évora, which may explain their
appearance.   Although he had to lecture in Latin, and was
writing and speaking Latin continuously, he shared
quarters with two scholars and fine poets, Resende and
Clenardo, who were very well versed in Latin, Greek and
Hebrew.   Moreover Nunes was one of Europe's leading
translators of Greek scientific works, by Ptolemy, Euclid
and Aristotle, the latter especially hard to translate.   His
head must have been full of Greek at that time, and his
early Greek poems seem to have poured out of it, at times
without sufficient attention to the repetition of phrases
and to exact word usage and accentuation.

The last nine are on the death of Prince Louis, and among
the earlier ones, nine are to members of the royal family,
like King John III (five), Cardinal Henry (one, see above),
Edward (one), Louis, while still alive (one), and the newly-
born son of Prince John, Sebastian (one).   As a member of
the Royal Court and tutor to the Princes, this large group
of royalist poems comes as no surprise.   Nor the equally
large group of religious poems (15), which combine with
the religious themes of most of his Greek poems on the
death of Louis, to suggest Pedro's strong Catholic fervour,
perhaps a defensive reaction to his own Jewish
background.   The only possible reference to this awkward
past is a four verse theme that appears on no less than four

occasions, in which the land of the Jews is described as the ideal land for one's forefathers to have inhabited.[100]

Of the fifteen poems which are religious, twelve are on Saints: Francis (three), Jerome (five), Remigius (two), Peter (one) and Antony the Hermit (one), while two are on the Eucharist and one is on Death. There is one also on the city of Lisbon, at that time one of the busiest and most colourful harbour-cities in the world, its port crammed with caravels and merchant ships that were bringing in the spices and silks from the East to suit the tastes of Northern Europe. At the end, Nunes argues that King John III's virtue has triumphed over time, which was wearing away the fine buildings and ships of Lisbon. The eulogy is brief, but worth quoting (my translation):

On Lisbon

This city is fit for all men to live in, with its weapons and
    its wealth, a city famous for battle-ready people,
celebrated for the beauty of its houses, for its citadels,
    for its very bounteous ocean and its broad river.
It has ships amazing in their size and number, that are   5
    guarded by its beautiful and famous harbour.
Time wears all of these away, and thus the virtue of your
    King has truly brought you more and greater honours.

Finally, two of the poems are on tertiary education, as one might expect from a first-class University teacher and tutor. In them Grammar is given a very high rating, just as it was in André de Resende's *Oratio pro Rostris*, delivered before the professors of the University of Lisbon on October 1, 1534, and published very soon afterwards, dedicated to King John III.[101] In my English translation the two poems read as follows:

On Grammar

Train yourself in Grammar, the nursemaid for all of the
    sciences, never ceasing from hard labour, and if

you should chance to use it for the complex puzzles of the sciences,[102]you will reach the desired light with joy.

On the same Grammar

Grammar gives many rewards to all of the sciences,
    alone having a beauty unknown to many men.
Rhetoric provides bright splendour and every richness for
    others, even if they are unwilling to study it properly.
Beguiling Poetry soaks our eyes with a strange beauty   5
    while uttering a gladdening melody in our ears.
Logic, astute where misleading complexities are involved,
    deceives one, concealing itself in a fog of obscurity.
Only Grammar, that rightly detests the others' riches,
    is itself exceedingly rich, in reality.   10

The stress on hard work again suggests that Pedro Nunes was in fact a very demanding teacher, perhaps more so than the young Pinheiro. After lecturing on Rhetoric[103] and Logic at the University of Lisbon, he is happy to stress their importance, although the obscurity of Logic is somewhat unflattering. It seems likely that Nunes would have used the Κατηγορίαι (*Categoriae*) and Τοπικά (*Topica*) of Aristotle for his classes, rather than the simplistic explanation of these works in the *Isagoge* of Porphyry (c. 232-c.306 AD), the usual text for Medieval and Renaissance classes studying Logic.[104]

The charm of poetry was all around him, with Resende and Clenardo busy composing Latin poems, but from start to end he argues for the greater importance of a sound grasp of grammar. The first Latin grammar in Portuguese was in fact composed by his fellow Royal tutor, André de Resende, while he and Nunes were teaching at Court, and it was published a few years later, in 1540.[105] In it Resende criticized the sloppy grammar and distorted speech of the Courtiers of his day, in a particularly interesting passage with regard to the language disputes of the sixteenth century. Written very grammatically, it reads as follows:

the title *De Arte atque Ratione Navigandi.* Both Resende and Nunes were equally fluent in the two languages, at a time when Latin was the norm, whether lecturing at the University, or acting as tutor to the royal Princes. In fact it was the only language allowed by the King. Both of them helped considerably in the establishment of a literary language for Portuguese prose.

Finally, the comparison between Nunes and Francesco Filelfo (1398-1481) mentioned in the Introduction, is enlightening and worth closer attention, even though Filelfo was Italian and had lived a century earlier. For longevity, his 83 years surpassed Nunes' 76 years and Resende's 75, despite enduring two stays in prison, two assassination attempts, several periods of near starvation, and many close calls with plagues and inter-city wars. In fact, by Renaissance standards, all three humanists were surprisingly long-lived. More important, however, is the fact that both Filelfo in Italy, and Nunes in Portugal, were pioneers in writing Ancient Greek poetry.[107] The only major works in Greek poetry in Italy after Filelfo were Poliziano's Greek epigrams, and, as shown in chapter 3, Nunes was the only humanist of his day in Iberia to attempt original poetry in such a difficult medium.

Both poets made use of Attic and Homeric language and prosody, both borrowed stock formulae (in excess in places), both showed little real originality, both wrote a substantial body of poems (44 in Filelfo's *Psychagogia*, and 36 in Nunes' *opus*), both chose the epigram (alternating with Sapphics in Filelfo's *opus*), and in both the subject-matter was either philosophical/religious, or literary, or encomia dedicated to princely/papal patrons, who were in fact far more generous and reliable for Nunes than for the hard-up and ever-complaining Filelfo.

Chapter   Eight

Religious   Commentary

Unlike his early *álgebra*, the following religious notes, mainly on the Resurrection and man's pilgrimage, on Jesus' miracle of the loaves and fishes and on the Annunciation, were scribbled down by Nunes in Spanish, with many abbreviations and without full attention to spelling. This made them very hard to decipher. In several places Nunes wrote down a word and then crossed it out for a better alternative, or left a blank, which shows that this was his rough copy, not in any way designed for publication, a factor that gives it greater interest. It seems more than likely that these comments were jotted down in the few still empty folios of the teacher's note-book that he had used in teaching the Princes in Évora, and they were random religious thoughts for his personal use only. A similar personal note-book for his religious thoughts was kept by Prince Edward, and only discovered in his bed after his premature death. Unfortunately it has not survived.[108]

At the time (1563), Nunes was finalizing his full-scale Spanish version of his *álgebra* for publication in Antwerp, translating and filling out the Portuguese original. He must have been fluent in Spanish ever since his early days in Salamanca, where he had studied at the University and had met and married his Spanish wife.[109] His quotations and comments are as follows, each section presented first in the original Spanish (except for his Latin quotations), and then in English. The script is very hard to read, and there are a few words that cannot be deciphered with certainty. These are followed by question marks.

A. *Ecce duo ibant* (Luke 24) *Prima octava Resurrectionis.*

Solus peregrinus ad Hierusalem.
El hijo de Dios vino del çielo a enseñar nuestra ignorançia y dar socorro a nuestra flaqueza. Lo primeiro hizo con su doctrina; lo segundo con su graçia. Entre las otras cosas que nos enseñó fue ésta mui prinçipal - a bivir como peregrinos.

1. Por esto naçió como peregrino en casa ajena *sc.* el pesebre. Bivió como peregrino. (Matt. c. 8.) *Vulpes foreas habent .... reclinet.* Murió como peregrino en casa estrecha en la cruz. Fue enterrado como peregrino en el monumento de Nicodemus.[110] Después de resusçitado estos 40 días que sólo anda peregrinando, mas aún paresçe aí en espeçie de peregrino.

2. La razón desto es porque el açierto de los hombres está en tomar cada cosa como ella e. Y la vida es camino o peregrinaçión o navegaçión. Metéis [os]'[111] en una nao, hecháisos a dormir; el agua os lleva sin sentirlo, de manera que muchas vezes os dizen que estáis en el puerto. Así es el hombre, que va metido en este tiempo como en una nao; y así lo pasado no paresçe nada. Y así, Genesis 47, respondió Iacob a Faraón: *Dies peregrinationis meae 130 ... meorum..'* Y nunca torna el hombre a arrib[ar].

3. También lo que nota bien S. Basilio. *Psalmo 1. In via peccatorum non stetit.* Los que caminan, el primero da su paso, y luego el otro y en continente el que viene tras él. Considerad un poco y veréis lo mismo en la vida: oi labráis vuestra heredad, daquí a un poco la labrará otro, y después de un poco otro. Veis unas casas magníficas, preguntáis cúias son; dízenos de tal rei. Y poco antes? De su padre, etc.

4. Y lo que más es para notar: que el caminante todo lo toma como de camino. Si topa un lugar mui fresco, alégrase

un poco y pasa por delante; y de aí a un poco topa con camino de sierras y síntelo poco. Así es la vida, que ni sus alegrías duran, ni sus miserias se acaban. En un soplo se acabó la prisión de Ioseph y en otro su privança con Faraón, en un soplo se acabaron las persecuçiones de David y en otro su reino majestad temporal y descanso, en un soplo se acabó la adversidad de Mordocheo y en otro su alegría y prosperidad. Esto luego nos enseña Cristo: pasar por todo como peregrinos. Los quales si son de alguna tierra noble todo lo que tienen en ella algún hombre poderoso, son [sin?] esta[s?][112]

[l.h. margin] Hierem. 14: *quare sicut colonus futurus es in terra et quasi sicut viator declinans ad manendum.* (Genesis 46) *Iam laetus moriar quia vidi faciem tuam, et superstitem te reliqui.*

- Jesus -   1563

A. "And behold two of them went" (Luke 24,13) The first episodes of the Resurrection. "Thou art a pilgrim going on thine own to Jerusalem." The Son of God came down from Heaven to enlighten our ignorance and to give succour to our weakness. The former He achieved through His teaching, the latter through His grace. Amongst the other things that He taught us was this most essential thing: to live as pilgrims.

1. For this reason He was born a pilgrim in a strange house, *i.e.* the manger. He lived as a pilgrim. (Matthew 8, 20): "Foxes have holes ... lay his head". He died as a pilgrim in a poor house, on the Cross. He was buried as a pilgrim in the tomb of Nicodemus. After His return from the dead, during these forty days, He not only goes about as a pilgrim; but even appears there in the likeness of a pilgrim.

2. The reason for this is that the right choice for humankind is to take each thing as it comes. And life is a journey or a pilgrimage or a voyage. You board ship and

lie down to sleep. The water carries you without your feeling it, so that they often have to tell you that you have reached port. So it is that humankind is on board time, as if on a boat, so that what has happened seems to be nothing. So then, Genesis 47, Jacob replied to Pharoah: "The days of the pilgrimage of my life 130 ..... of my [fathers]". And man never returns to the port [of departure]..

3. Furthermore, as St Basil observes on Psalm 1 "[Blessed is the man... nor] standeth in the way of sinners", when people walk, first one takes a step, then another and straightaway the next person follows suit. If you think about it, you will see the same thing in life: today you labour on your inheritance, and a little later someone else will be doing so, and soon afterwards yet another. You see some splendid houses and ask to whom they belong. They tell you "to King so-and-so". And before that? "To his father", and so on.

4. It is most noteworthy that the traveller accepts everything as part of the journey. If he comes to a cool, refreshing place, he takes his ease for a while before going on his way; a bit later on he reaches a rough, mountainous path, and hardly notices it. So it is in life, for neither are its joys lasting, nor are its sufferings without end. In a single moment[113] Joseph's imprisonment ended, and in another, his position as Pharoah's adviser. In a single moment the persecution of David ceased, and in another, his temporal power and peace of mind. In one moment the adversities of Mordechai came to an end, and in another his happiness and prosperity. This then is what Christ teaches us: to pass through all things as pilgrims. They, if they are from some noble country, should be humble(?), Jeremiah 14, "wherefore just as you shall be a settler on earth, and just like a traveller turning aside to pass the night." Genesis 46.30, "Now let me die, since I have seen thy face, and leave you still alive."

B. 1. Lo segundo que es para considerar[114]: por qué razón Cristo Nuestro Señor, para hazer este tan señalado milagro como es hartar 5,000 personas, pide 5 panes de çevada, pues hazía tan poco al caso tan poco pan para tan poca[115] gente. Y así, Matt. 15, para 4,000 pidió los 7 panes y en el conbite de las bodas para proveer de vino hizo henchir de agua las tinajas.

2. La razón desto es por que quiere Dios que, para reçebir dél grandes mercedes, a lo menos pongamos de nuestra parte algún pequeño de[116] trabajo. Avéis ofendido a Dios. Metéisos en vuestro estudio; pensáis la fealdad del pecado y la pena que mereçéis; lloráis por temor desta pena unas pocas de lágrimas. Ya hechastes agua en los vasos. Esperad que Dios las convierta en vino, del çielo dando su graçia. Venís a la confesión tibio, mas enfín venís. Con diligente examen y con un dolor imperfecto, pan traéis, mas de çevada. Esperad, que Dios con el calor del sacramento os dará contriçión. Y por eso es bueno en el vaso del entendimiento hechar cada día aunque no sea sino una gota de consideraçión de la muerte, en la voluntad una gota de deseo de la gloria, por que Dios hinche de fuego.

3. Así aconteçió (IV Reg. c.4) *Mulier autem quaedam de uxoribus prophetarum ..... stetitque oleum.*

3. i. Lo último que ai para notar que esta gente no se levantó con lo que sobró, antes los 12 compañeros del Señor recogieron 12 alcofas,[117] [por] lo qual San Crysóstomo encareçe con razón la providençia de Christo y saber, en tener tan tasada la hambre de toda esta gente que quedasen 12 cestas para que cada Apostolo recogiese la suia. La paçiençia Christi dexa que Iudas también recogiese la suia.

3. ii. Daquí finalmente comiençan a manifestarse las razones por las quales Dios Nuestro Señor da estos tiempos tan estrechos y á tanto que duran; porque esta gente salió a buscar a Cristo Nuestro Señor al desierto y, salíendo a oír la doctrina del çielo, se olvidava de la sustentaçión de la tierra; y así Cristo le dio una y otra. El mundo anda todo

metido en la sustentación del cuerpo y tiene mui poco que ver con la del çielo, y así pierde una y otra. Como quando buscamos una cosa, el andar sobre ella mucho y mui embebeçidos es causa de no verla perderla, assí el embebeçimiento en las cosas temporales. Esta gente como nota asentóse sobre el heno y pisólo, que como nota San Hierónimo es pisar los deleites, el cuerpo, la buena vida, porque todo eso es heno: *omnis caro faenum* .[118] El mundo dase al contrario, que es poner el heno sobre la cabeça, buscar los deleites por mar y por tierra, y así viene Dios y cierra la mano de la madre que toma la llave.

3. iii. Esta gente asentóse por orden y reçibió su sustentaçión por orden de los discípulos del Señor, que nota San Lucas [9. 14-15] mui bien. Y es que las mercedes de Dios las reçibimos por mano de nuestros prelados, y así donde viésemos un prelado y aún un sacerdote, avíamos de tener altísima reverençia. Agora es entrado en el mundo por nuestros pecados grande despreçio de los saçerdotes y aún de los prelados, siendo tanto por el contrario en el tiempo antiguo que Constantino Magno, como se refiere en el primero tomo de los *Conçilios* , señor del mundo, en aquel grande Conçilio Niceno de 318[119] obispos por reverençia no se asentó entre ellos, y quando se asentó fue en un escabello más baxo. Y por ver Dios que no bivimos etc.

3. iv. Esta gente dio lo sobrado; el mundo, quando ai buenos años y que sobra, levántanse[120] con ello. Viene Dios y quita los temporales, porque son plaço, que se pierde quando no se acuerde con el fuero. Y éste es: dais a los pobres.

3. v.La razón que tenemos es grande porque emos reçebido tantos limosnas: la vida, la sustentaçión, el sacramento, el mismo Dios. Y finalmente a trueco desta limosna emos de reçebir aquella bienaventurada limosna del reino de Dios.

Venite benedicti patris mei ad me. Laus Deo.

B. 1. The second thing is to consider why Christ Our Lord, in order to perform such an outstanding miracle as the feeding of the 5,000 people, asks for five barley loaves, since so little bread could hardly meet the needs of so many people. Thus, Matthew 15, He asked for seven loaves for 4000 people, and at the wedding feast, to provide enough wine, he had the wine-jars filled with water.

2. The reason for this is that God wishes us to make at least a little effort on our part in order to receive His bounty. You have offended against God. Then you go into your study, you think about the ugliness of your sin and the punishment you deserve and for fear of that punishment, you weep a few tears. Now you have poured water into the glasses; wait for God to turn your tears into wine, offering His grace from Heaven. You come to confession half-heartedly, but at least you have come. With careful soul-searching and with a still limited repentance, you offer bread, but of barley meal. Wait, and God will give you contrition, through the warmth of his sacrament. And for that reason it is a good thing to let fall each day into the cup of understanding, even though it be no more than a single drop of the contemplation of death, and into the will, a drop of the desire for [heavenly] glory, so that God may fill you with his fire.

3. Thus it happened that (II Kings ch.4 [1-6]) "[Now there cried] a certain woman of the wives of the sons of the prophets [unto Elisha] ...... and the oil stayed."

3. i. The last thing to be noted is that these people did not carry off what was left over, but rather the twelve disciples of Our Lord gathered up the twelve baskets full. For this reason, St Chrysostom rightly praises the providence and wisdom of Christ, in knowing so perfectly the hunger of all these people that twelve baskets full should remain, so that each Apostle could gather up his own portion. The forbearance of Christ even allowed Judas to gather up his.

3. ii. Finally, in these things the reasons begin to become clear why Our Lord God gives us such hard times as these, and why they last so long. Because these people went into the desert in search of Christ Our Lord, and going out to hear the doctrine of Heaven, they forgot earthly sustenance, therefore Christ gave them both the one and the other. The world, too much involved in bodily sustenance, has little time for Heavenly food, and so it loses both. Just as when we search for something, walking over it while wrapped up in other matters causes us not to see it, even so, being wrapped up in temporal affairs [stops us].

3. iii. These people sat upon the grass and trod it down, and this, as St Jerome observed, is to tread down pleasures, the flesh, high life, for all of that is grass - "all flesh is grass." The world gives itself over to contrariness, covering its head with grass, seeking pleasures on land and sea, and so God comes and closes the hand of the mother who picks up the key.

3. iv. These people sat themselves down in orderly fashion and received their sustenance by the orders of the disciples of Our Lord, as St Luke [9. 14-15] notes very clearly. And it is a fact that we receive God's favours from the hand of our prelates, so that wherever we saw a prelate, or even a priest, we should have shown him the highest reverence. Nowadays, because of our sins, there has come into the world a great disrespect for priests and even for prelates, which was quite the opposite in days of old, when Constantine the Great, the master of the world (as is related in Book 1 of the *Councils*), in the great Nicene Council of 318 bishops, through reverence for them, would not sit down among them, and when he eventually did sit down, sat on a lower stool. And since God sees that we do not live [etc[121]].

3. iv. These people gave up what was left over. The world, when there are good years and there is enough and to spare, builds itself up with it. God comes and takes away the temporal goods, because they are a pledge that is lost

when it is not in accord with our pledge. And that is "Give to the poor".

3. v. The reason we have is a great one, for we have received so many benefits, our life and sustenance, the sacrament and God Himself. And finally, in place of these, we are to receive that blessed gift of the kingdom of God. "Come to me, ye blessed of my father. Praise be to God".

C. *Missus est angelus ... In Annuntiatione* 25 Martii[122]

1. *Missus est angelus ... Maria* : Ase de notar que el camino que vuo en nuestra perdiçión, ése convirtió Dios en el contrario para nuestra salvaçión.

Demonio, serpiente, muger, plática, creer y obedeçer al mal. Dios, ángel, muger, plática, creer y obedeçer al bien.

2. *Sic gloriosus artifex quod quassatum erat non infregit, sed utilius refecit* :

Así aconteçe si tenéis un anillo que estimáis, si se quiebra, no lo deshazéis, antes procuráis que se suelde, y en aquel lugar se ponga una rubí. Así, para soldar la perdiçión nuestra, lo hizo Dios que no nos destruyó, antes soldó el mal complido por Eva con María, y puso en ella aquella piedra preçiosa del çielo engastonado en su carne.

3. *Simile est regnum caelorum homini .... illum* :

Desto tenemos una historia en la Sagrada Escritura en Esther, donde leemos que la pena que Amán avía ordenado para Mordocheo, ésa padeçió, y la honrra que para sí quería, ésa fuera dada a Mordocheo. Amán significa el demonio, el qual sufre[123] la confusión que ordenava al hombre, y el hombre tiene la honrra que quería para sí: *ero similis Altissimo.*
Sacamos este provecho: que por los medios con que nos pedimos, al contrario nos ganamos. Exemplos: al murmurar

suçeda dezir bien; al distrahimento, recogimiento. Ioan. *humani dico ..aeterna.*

4. i. *Et ingressus angelus ...in mulieribus* :
Esta es la condiçión de Dios, que quando los hombres esconden mucho sus loores, enbía ángeles a publicarlos. Los maiores loores de pura criatura fueron los desta Santa, los más escondidos por su dueño. Qué haze Dios? Házelos cantar por los ángeles.

4. ii. Avía Christo Nuestro Señor estado 30 años sin se manifestar. Quando vino a manifestarse, fue escondiéndose y suietándose. En el mismo auto viene el pregón del çielo: *Hic est filius meus dilectus ... complacui* . Aún después desto la Virgen gloriosa para montar (?) más antes a lo que después avía de hazer su hijo, llena ya de Dios, va a servir a Elizabeth, y allí mesmo el Spíritu Santo despierta la lengua ançiana en sus loores. *Beata... Domino* : y contentó esto tanto a Dios que avía hecho cantar esta propriedad desta Santa por David. *Omnis decor eius filiae regis admiratus ... regis* .124

4. iii. La razón desto es que quiere Dios las virtudes sin ser manoseadas y publicadas, porque a los reyes lo que se les pone delante á de ser cosa en que no toque la gente. De donde se ve el ierro del mundo, que cada uno es pregonero de sí mesmo, y no temen aquellas amenazas de David,125 que conoçía bien la condiçión de Dios. *Dominus ossa corporis .... sprevit eos*. Y el mismo David lo experimentó bien quando se descuidó dello contando el pueblo.

C . "An Angel was sent" Feast of the Annunciation, 25 March

1. "And an Angel (Gabriel) was sent ... (name was) Mary".
It should be noted that the path that led to our perdition was changed by God into its opposite, for our salvation.

Devil, Serpent, woman, talking, believing and obeying evil.

God, Angel, woman, talking, believing and obeying good.

2. "Thus the glorious artist did not break up what was damaged, but repaired it to be more useful." It happens like this if you have a ring that you value. If it gets broken, you do not destroy it, but try to have it repaired, and fix a ruby where the join is made. In the same way, to repair our perdition, God acted in such a way as not to destroy us, but rather to repair through Mary the evil committed by Eve, and he set in her that Heavenly jewel mounted in her flesh.

3. [For the kingdom of Heaven] is like [unto a man] ...[126] We have a story about this in Holy Scripture, in the Book of Esther, where we read that the punishment ordered by Haman for Mordecai, he suffered himself, and that the honour which he sought for himself was given to Mordecai.[127] Haman stands for the devil, who suffers the confounding that he had planned for humankind, while humankind holds the honour that he sought for himself; "I shall be like unto the Highest". We learn this lesson: that through the opposite of the means whereby we were lost, we may regain ourselves. For example: for speaking evil, substitute praise; for distraction, concentration. John. "I call human .... eternity".

4. i. "And the angel came in [unto her] ... among women."[128] This is God's way, that when people take care to conceal their own praiseworthiness, He sends angels to publish it abroad. The greatest causes of praise for a pure creature belonged to this holy woman, and were most concealed by her. What does God do? He has them sung by angels.

4. ii. Christ Our Lord had spent 30 years of his life without revealing Himself. When He came to the point of revealing Himself, it was by hiding Himself away and by subjecting Himself. In this very act came the proclamation from Heaven "This is my beloved Son, in whom I am well pleased." And so after this,[129] the glorious Virgin, so as to rise nearer to what her Son was to do later

on, already heavy with God, goes off to wait upon Elizabeth, and in that very place the Holy Spirit awakened the old man's tongue in praise of her: "Blessed [art thou among women] .... from the Lord".[130]   And this greatly pleased God, who had caused this property of this holy woman to be sung by David. "All the beauty of that daughter of the King, admired ..... of the King."

4. iii.   The reason for this is that God desires virtues that are undefiled and unpublicized, because what is set before kings should be something that is not touched by the common people.   From this we can see the error of the world, in which everyone sings his own praises, and has no fear of those warnings of David, who well knew God's disposition. "The Lord .... the bones of his body .... despised them."   And David experienced this himself to the full, when he disregarded it by counting the people.

D.   Lo que conviene luego es hazer bien y dar toda la gloria a Dios, porque mientras más se atribuien todas las cosas a este Señor, mas cuidado tiene Él de acudir con la recompensa.

1. *Quae cum audisset ...... Deum* : Los que han perdido alguna cosa ponen gran diligençia en saber quién la halló? Nosotros avíamos perdido la graçia;   pues la Virgen la halló, vámonos a ella. Y con la graçia el alegría, las virtudes, la esperança. Todos luego se vaian camino a la Virgen, porque ella no sólo halló la graçia, que es semejança de Dios, mas es llena de graçia.

2.   Y por eso la Escritura Santa unas vezes la llama 'luna', otras 'sol', otras 'mañana', porque así como viniendo la mañana huien las bestias fieras y se esconden, y las enfermedades se alivian, y los hombres si tienen más esfuerço, así viniendo esta mañana aconteçió todo esto, y todo el tiempo que preçedió la Virgen fue noche escura. [Cant. 6]

3. Quando Jacob, Gen.32, tornava a su tierra, luchó toda la noche con el ángel y no alcançó su rendición sino quando vino la mañan. Y de Jacob, que quiere dezir hombre que lucha, fue llamado Israel, hombre que ve a Dios. Así mismo, los hombres trabajaron[131] todo el tiempo que preçedió la venida de la Virgen por alcançar perdón de Dios. Los ....[132]

D. What we should do, therefore, is to do good and to render the glory to God, because the more all things are attributed to this Lord, the more pains He takes to hasten to reward us.

1. "When he had heard these things, ... God"
Those who have lost something take great pains to discover who has found it. We had lost grace; since the Virgin found it, let us turn to her. And with grace, happiness, the virtues and hope. Let all now turn towards the Virgin, for not only did she find grace, which is a likeness of God, but she is full of grace.

2. And for that reason the Holy Scripture sometimes call her the Moon, and at other times the Sun, and at other times the Morning. For just as with the coming of morning, the wild beasts run away and hide themselves, and sickness eases, and man regains his strength, even so, when this morning came, all this came to pass, and all the time before the Virgin came was [like] dark night.

3. When Jacob, Gen. 32. 24-28, was returning to his homeland, he wrestled with the angel all through the night, and he could not force him to surrender until morning came. And instead of 'Jacob', meaning "the man who wrestles", he was called 'Israel', "the man who sees God". In the same way, all through the time before the coming of the Virgin, men laboured to obtain God's pardon. The ....

At this point the Spanish text breaks off. Further sheets may have been filled with his musings, but have long since dropped out. Enough remains, however, to give us a good idea of the very Christian piety of the writer, with his special love of the Blessed Virgin Mary. His quotations appear to be from memory, and cover both the Old and the New Testaments, clearly very well known by him, plus the works of Chrysostom. His quotations are well used to illustrate his admonitions and exhortations - made for his own benefit, it seems. He confused Nicodemus with Joseph of Aramathea, but otherwise made no biblical mistakes, despite such a wide range of Latin quotations.

Like the Meditations of the Roman Emperor, Marcus Aurelius, these homilies are still very valid today, especially their stress on the need for humility and for a proper understanding of the brevity and uncertainty of human life, their demand for proper reverence for the clergy, who were suffering from disrespect in Nunes' day (as now), and for generosity to the poor. Heavenly grace and happiness, he assures himself, will come from resisting pleasures, from regular communions, from doing good deeds and from glorifying God, but most of all, through the love of Christ and his Holy Mother. Nunes' traditional contrast between the prototype temptress, Eve, and the Virgin Mary, who redeems her sins, is Mediaeval in its outlook. It reveals the male dominance in the Church affairs of his day, which has continued since then almost without question in the land of his birth.

The opening section on man's journey through life suggests that he was in his later years, nearing the end of his own pilgrimage, and soon to reach his final port. The nautical imagery in A(2) is perhaps surprising, if he never went to sea, but not when one considers how much of his life he spent with Portuguese mariners. The idea of boarding ship and arriving at one's destination without noticing the water slipping beneath one is aptly used for the ever faster passage of years.

Written by Nunes, these meditations put an end to the idea that he ever returned to his Jewish background. The heartfelt devotion to the Catholic Church, to its clergy and especially to the mother of God, suggest a true believer, a strict adherent to his country's religion and an avid reader of the Bible and of Church history. This was the last chapter in Nunes' extraordinary life, its richness now shown far more clearly by the contents of this unpretentious manuscript from Évora's public library.

Appendix A

Pedro Nunes: Tables of Contents

LIVRO DO ÁLGEBRA(1533) LIBRO DE ALGEBRA(1567)

PRIMA PARTE

Parte   Secunda

Parte   Tertia

TERTIA PARTE

Translation

The Portuguese for the chapter headings can be seen in English in the text of the Algebra, but for convenience they are set out below, as follows:

Coniugações = Conjunctions
Denominações = Denominations
Asomar Inteiros = Adding integers
Deminuir = Subtraction
Multiplicar = Multiplication
Partir = Division
Reduzir quebrados = Reduction of fractions
Abreviar = Condensing fractions
Das raizes = Roots
Regra geral ao asomar raizes = General rule for adding roots Proporções = Proportions
Igualmentos = Equalization
Problemas Compostas = Problems with compound conjunctions.

The Spanish equivalents are too similar to the Portuguese to need translation.

Appendix B

Anonymous Latin Poem on
A Dama da Cutilada

The dramatic event described in the mock-epic poem below was associated with Dona Guiomar, the daughter of Pedro Nunes. It took place in Coimbra on Friday 17th January, 1578, only seven months before the mathematician's death. It is quite possible that the shock and disgrace of her violent act accelerated his demise.[133]

The tradition is that this Guiomar was living with her retired father at Calçada in Coimbra, and was engaged to be married to a noble young neighbour, Heitor de Sá, whose family appears to have opposed his union with Nunes' well-off but plebeian family. Solemn oaths had been exchanged, but Guiomar's fiancé not only left Coimbra, but soon became engaged to another lady. Pedro Nunes complained to his old friend, Manuel de Meneses, Bishop of Coimbra, who gave orders for Guiomar and Heitor to appear before him together with their families, at the Church of St John of Almedina. Local dignitaries also attended, and Heitor was duly criticized for his behaviour, but in his defence he argued that he had never in fact promised to marry her.

Guiomar's own pride and sense of honour and her feeling of helplessness before the Law, combined with her anger and her hot Spanish blood, led her to produce a knife from under her dress, and with this weapon she slashed the face of her handsome betrayer, marking him so badly that no other woman would ever want to look at him. For the beautiful Guiomar, vengeance was sweet, but she was at once arrested by the Bishop and thrown into his prison. However, two months later, on 21 March, she was smuggled out in a large basket and taken to the convent of Santa

Clara, where she became a nun. This left Nunes without a daughter at home, as his first two daughters, Isabel and Briolanja, had left home to be married, respectively, to a João Pereira de Sampaio[134] and a Manuel de Gama Lobo, while his youngest daughter, Francisca, was a nun in Lorvão. Both of his sons were abroad, serving in India, and were possibly already dead, which left the elderly Pedro and his wife very much without support.

The story of Guiomar was very popular with poets and novelists, both at the time of her attack and in later centuries. The mock-epic poem below seems contemporary.

De Conimbricensis Hectoris memorando casu

Horrendum dictu facinus per saecula numquam
auditum offertur, superat quod facta virorum
incluta nobilium, veteres renovatque triumphos.
Est novus in terris generoso a sanguine cretus
Hector, inhumanas ferro qui vincere gentes, 5
terribilis, Maurosque feros et regna domare
indomita, et Turcas ferro trucidare feroces
sanguineo pictosque Indos submittere regi
ipse suo, gelidosque Scythas acresque Liburnos
sternere, et Europam et populos pacare calentes, 10
ignibus assiduis gentes frenare superbas,
iactabat se posse feras superare phalanges.
Musa, precor, coeptis faveas. Conimbrica tantum
Hectora donavit terris, qui captus amor
virginis, indomito infelix sic arserat igni 15
ut nova continuo tentaret furta puellae.
Et nunc attonitos figens in virgine vultus,
haeret inexpletum, noctemque diemque fatigat
questibus assiduis, curas partitur in omnes
infelix animum. Sensus nunc carmine mulcet 20
virgineos, citharaque sonans fidibusque canoris
securos insomnis amat perrumpere somnos.
Et modo purpureis intexens lilia demens
alba rosis,[135] *capiti donat gestanda puellae*
munera, sed mentem primaevo flore pudicam 25
nulla movent mandata; preces volvuntur inanes.
At miser assiduis cruciatibus actus amoris,
ut vidit sua vota vagos ferre improba ventos,
coniugium exposcit, membris ut ferret amica
otia defessis. Haec cum manifesta propinquis 30
essent, extemplo demens simulatur ab ipsis.
Iam manicis artare manus, in vincula duci
dura iubent, insignem aegre graviterque ferentes
connubio iungi tali. Iam cuncta negare
acta monent, iam vincla viro manicaeque levantur 35
Ille acri devectus equo, loca multa pererrat,
insignis torque aurato croceoque galero
vesteque purpurea, quo cuncti mente putarent

insana iuvenem veros agitare furores.
At coniunx subito mentem percussa dolore, 40
in varias animum curas deducitur anceps
an miseram immatura petat per vulnera mortem,
an repetat demens meritas a sanguine poenas.
Ergo ad iudicium ductus cum coniuge, iussu
praesulis, audaci rem totam fronte negavit, 45
contendens numquam taedas celebrasse iugales.
Ast illa incautum, ferro quod veste tegebat,
occupat, et medios inhonesto vulnere vultus
dividit. Exsuperat calidus percussa repente
ora cruor tepidique sinus fuso imbre madescunt. 50
Obstupuit praesul. Iamiam fera bella propinqui
intendunt. Nunc, Musa, graves mihi suggere motus,
ingentes iras animosque ad bella paratos.
Horrendis nam quisque parat cruciatibus illam
torquere, ast alii laqueum, pars dura minantur 55
supplicia, exclamant alii trepidantque furentes.
Ut solet in populo causis cum saepe coorta est
seditio levibus, rabidus furor arma ministrat,
iamque faces iacit incensas ignobile vulgus
viribus immensis, liquidumque per aera saxa 60
dura volant; rabies animos immensa feroces
impellit. Tum forte virum pietate verendum
si videre, tacent arrectisque auribus astant.
Idem terribiles animos et pectora sedat.[136]
Humanus monitis praesul sic temperat iras 65
indomitas, animosque viri praesentia flectit,
accipit et placide manicas animosa virago.
I decus eximium, meritas adipiscere laudes,
quae vere illustras maiorum facta. Manebit
femineaeque manus signum tibi vulneris Hector. 70

6 *Mavors* T[1]
63 *et promptis auribus* T[2] *in margine*

On the memorable fate of Hector of Coimbra

A crime terrible to relate and never heard of for ages
is offered here, which surpasses the famous deeds
of noble men, and one which recalls ancient triumphs.
A new Hector was born on earth of noble blood,
who used to boast that with his bloody sword he   5
could conquer savage races, striking terror in them,
and tame wild Moors and unconquered kingdoms,
and cut down ferocious Turks, and submit
coloured Indians to his King, and lay low the
frosty Scythians and fierce Liburnians, and   10
pacify Europe and sunny peoples, and control proud
races with assiduous fires, subduing fierce squadrons.
Muse, I pray you, favour my undertaking.   Coimbra gave
this great Hector to earth, who fell in love with a young
maiden, and burned with such uncontrollable passion,   15
poor wretch, that he at once tried to deceive her in a
new way.   And now fixing astonished looks at the virgin,
he clung to her incessantly, spending weary days and
nights with endless complaints, and wretchedly dividing
his mind among every care. He charms her maidenly   20
senses now with poetry, playing a lute and sonorous
strings, he loved to interrupt her quiet sleep, unsleeping
himself. Now madly weaving white lilies into red roses,
he gave them to his girl as presents to wear on her
head, but no requests accompanied by young flowers   25
moved her modest mind; prayers poured out in vain.
But the poor man, driven by ceaseless torments of love,
when he saw the fickle winds carry off his persistent
prayers, demanded marriage, to bring pleasant leisure
to his tired limbs.   But when his relatives discovered 30
this, they at once pretended he was mad.   They now gave
orders for his hands to be manacled, and for him to be
put in hard chains, bearing his union in such a marriage
with ill-grace and annoyance.   They warned him to deny
all his actions; then chains and manacles were removed. 35
Born on a spirited horse, he wandered through many
places, splendid in a golden necklace and a yellow hat
and a purple robe, so that all might think that real

madness plagued the young man, sent out of his mind.
But his wife, her mind struck by sudden grief,   40
was drawn into various worries, uncertain in her mind
whether to seek wretched death with premature wounds,
or madly seek a deserved punishment for shedding blood.
Led therefore to court with his wife, at the command of the
President, Hector denied the whole affair with brazen   45
face, arguing that he had never celebrated marriage rites.
But she attacked him unexpectedly with a sword, hidden
in her robe, and split his face down the middle with a
shameful wound. Hot blood suddenly spouted over his face,
so struck his hot breast soaked with a spreading shower. 50
The Governor was dumbfounded. Relatives already urged
fiercer wars. Now suggest to me, Muse, their strong
emotions, mighty anger and minds prepared for warfare.
For each man prepared to torture her with grim
torments, some threatened the noose, others   55
harsh punishments, and others exclaimed, as they
trembled with fury. As is usual when sedition arises
in a people, often for light reasons, mad fury
provides arms, and now the ignoble mob hurls blazing
torches with immense strength, and hard rocks fly   60
through the clear air; immense madness drives on
ferocious minds. Then if by chance they see a man
revered for his piety, they are silent and stand with
attentive ears. The same man settles fearsomel minds and
breasts.   Even so the kind Governor tempered their   65
uncontrolled rage with warnings, and the man's presence
changed their minds. The bold virago placidly accepted
her manacles. Go, splendid honour, obtain the praises you
deserve, as you truly depict the deeds of ancestors.   Hector
remains as your proof of a wound from a woman's hand.   70

Appendix C

Resende's Poem on the Death of Prince Edward

As well as the previously unknown poem below, Resende wrote a very moving account of Prince Edward's death in his biography of the Prince, the illegitimate son of King Manuel. Prince Edward was only 25 years old when he died, in a very pious manner, on 20 October 1540, after some ineffective blood-letting by the Court doctor, Dr Ximenes. Resende was his teacher and confidant, and a life-time friend of his family. His biographical *Vida do Infante D. Duarte* was first published in Lisbon in 1789, and included in José Pereira Tavares *André de Resende: Obras Portuguesas* (Lisboa, 1963) pp. 71-132. These vernacular works were to play a key rôle in the development of a fluent, euphonious prose-style in Portuguese, although Resende's Latin works had a far wider readership, and in northern Europe he was judged the prince of Latin poets. I hope to publish this biography, together with my English version and notes, in the near future.

Compared with the very Christian account of his death and burial in Resende's biography, the final picture in this poem of the Prince dining among Roman gods may seem somewhat incongruous. However, it suits Resende's picture in his biography of the Prince's extraordinary Classical erudition, his powers of memory and his admirable eloquence in Latin. As we have seen, his fellow royal tutor, Nunes, was equally prone to Classical imagery when describing the death of his beloved Prince Louis.

Exclamatio in Mortem Domini Odoardi

Umectate, viri lacrimis manantibus ora,
  guttaque per vestras vadat oborta genas.
Femina, quid cessas? Demissos solve capillos.
  Fle, puer, et longas solve, puella, comas.
Sol, propera retro terras de more patentes   5
  illuminans, nigras et redimito comas.
Luna soror, lucem spargens de nocte micante,
  desine longuinquam nunc remeasse viam.
Denique quaeque aulis vescuntur, pectora pugnis
  tundant, et nigra corpora veste tegant.   10
Mortuus est princeps, regum de sanguine cretus,
  nomine vulgaris, magnificus titulis.
Cui precor, alme pater, menses accumbere divum
  des, et pro meritis, ut reor ipse, dabis.

Exclamation over the Death of Prince Edward

Weep, men, with tears flowing down your faces;
  let tears spring up and cover your cheeks.
Woman, why hang back? Untie your hair, let it fall.
  Weep, boy, and untie your long tresses, girl.
Sun, hurry back from illuminating the open lands,   5
  as usual, and let your hair flow black.
Sister Moon, sprinkling light from a glittering night,
  now stop going back along your lengthy journey.
Finally, let any women live at Court   beat their breasts
  with fists, clad their bodies with black robes.   10
A Prince is dead, born from the blood of kings, common
  in his name, but magnificent in his titles.
I pray you, bounteous Father, let him feed with the Gods
  and you will grant what he deserves, I am sure.

Appendix D

Cardoso's Letters to Pedro Nunes and António Pinheiro

These flattering letters can be seen on folios 19 and 63-66 in the as yet unedited collection of letters written by, and in a few cases to, that eminent Portuguese humanist and pioneer lexicographer, Jerónimo Cardoso. He was born early in the sixteenth century in Lamego, and after obtaining his B.A. in Canon Law he became one of the professors in the Humanities at the University of Lisbon, at about the same time as Pedro Nunes, and in later years he also became one of the tutors of Prince Sebastian. Like Resende, he was given the honour of delivering a public oration at the opening ceremony of the University of Lisbon, in 1536.[137] There he distinguished himself for his success as a teacher, publishing a useful guide-book for his students, and for his ability as a Latin poet; to these can be added his attractive epistolary style, and his pioneer treatises on Portuguese and Latin grammar. His highly eulogistic Latin poem *Ad Olysipponensis Academiae Doctores* , included by him in his *Sylvae*, was written in 1536, but not published until 1564, 28 years later - well after the death of John III. As we have seen, the King had not been amused at all by the petition signed by most of the Lisbon professors, who were strongly opposed to the University's move to Coimbra. The signature of Nunes was not included, the issue being settled by the time he took over the chairs at Lisbon's University for Moral Philosophy, Logic and Metaphysics. The lectures in the final folios of his notebook manuscript,[138] give us a good idea of how he tackled the problem of teaching the art of public speaking to his students, almost all of whom must have been studying Law.

Clenardo's major work was his reverse Latin-Portuguese Dictionary (*Dictionarium Latino-Lusitanicum et vice-versa*

Lusitano-Latinum), the country's first such dictionary, published in Coimbra in 1569 and re-edited no less than seven times by the year 1694. His Latin poems appeared as two books of *Elegiae* and one of *Sylvae*, and he published a Latin treatise on the weights and measures of Greece and Rome. He died in Lisbon in 1569. One of his pupils was the celebrated humanist and historian, Jerónimo Osório, Bishop of Silves, who showed him life-long affection and great respect, as did many of his fellow teachers, especially Resende and Nunes, to judge from the letters below.

The first letter (on folio 19) was written to Pedro Nunes, and it can be dated to late in 1541, in the light of the reference to his recent Latin work.[139] Unfortunately, the reply by the eminent mathematician was not included by Cardoso. There is no modern edition of this very important collection of letters, and like almost all the others, this letter is undated. The Latin runs as follows:

Hieronymus Cardosus Petro Nonio Regio Mathematico[140] S. P. D.

Debere me humanitati tuae plurimum, Doctor eximie, haud quaquam diffitebor, qui vel tantulo accepto nuntio, libellum nuper latinitate donatum, quem magnopere lectitare gestiebam, e vestigio mittendum existimasti, ratus videlicet, id quod res habet, nihil tam mortalibus quam recentissima quaeque nec vulgo prius circumlata arridere. Quocirca haec cum diligenter perspexisses, sitim hanc flagrantem qua me opprimi sensisti sedare seu potius restinguere non distulisti. Unde coniectura ducor te litterarum studiosos, cum illarum arcem iam pridem tenueris, summo amore complecti. Itaque eundem libellum quem abs te in dies pauculos commodatum accepi remitto, ex quo voluptatem cepi plurimam multumque illius lectione sum oblectatus. Ceterum summi beneficii loco duxi tuam istam expromptam animi alacritatem, quam in mittendo ad me libello exhibuisti, polliceorque me <non> memorem numquam futurum, immo pro virili adnitar, te ipsum eruditionemque istam multijugam, qua polles et emines, posthac cumulatius evehere. Bene vale, emicantissimum doctrinarum omnium speculum, et me, si dignus videbor, non in postrema amicorum classe colloces pervellem. Vale.

Jerónimo Cardoso, greetings to Pedro Nunes, Royal Astrologer.

"Distinguished Doctor, I shall in no way deny that I owe a great deal to your kindness, for after receiving such a trifling message, you decided that your little book, recently presented in Latin, that I desired most eagerly to peruse, should be sent to me at once. You thought, of course, as was the case, that nothing pleased mortals as much as all the most recent and as yet uncirculated works. And so once you had carefully observed this, you did not delay in slaking, or rather in extinguishing, this burning thirst that you realized was tormenting me.

From this I am forced to conjecture that you embrace all students of literature with the greatest of affection, since you have long since held the highest position therein. And so I am sending back the same little book that I received from you on loan for just a few days, from which I derived immense pleasure and I was really delighted to read it. However, I thought that your ready promptness of mind provided the greatest benefit, shown by you in sending the little book to me, and I promise that I shall never forget it, rather I shall strive as best I can to broadcast hereafter more fully both you and that complex erudition of yours, wherein you are so well equipped and so eminent.

Farewell, most brilliant mirror of all learning, and if I shall appear to deserve it, I should like you hereafter not to place me in the lowest class of your friends. Adieu."

The second of Cardoso's letters is to António Pinheiro,[141] seeking his friendship, and highly praising his eloquence, perhaps partly due to his edition of Rome's greatest teacher of oratory, Quintilian (Book III) and his translation of Pliny's favourite oration, his *Panegyricus* in honour of Trajan. Again, Cardoso fails to include Pinheiro's reply. The Latin for his letter is below, followed by my English translation:

Hieronymus Cardosus Antonio Pinario Regio Contionatori S. P. D.

Etsi scio quam praepostere faciam, vir eminentissime, qui officium meum alio quam hoc tempore obeundum in hunc diem protraxerim, et scribendi ad te provinciam nunc denuo fuerim ingressus, tamen si verum ut incorruptus iudex examinabis, non id incuriae aut stuporis sed consilii potius et iudicii fuisse iudicaveris. Ego namque cum meae mihi imbecillitatis sim plane conscius, ut cui mediocria omnia (atque utinam tolerabilia) obtigere, veritus semper sum eos potissimum appellare aut scriptis meis lacessere, quos et fama consentiens et praeclara eruditio commendaret, ne aut Aesopicae ranulae exemplo rumperem, si me illis vel aequare vel conferre me studuissem, aut inopiam pannosque meos detegerem, dum vel copiam vel divitias ostentare decrevissem. Quod si aliquando ad viros eruditos mihi litteras dare contingit (ut certe contingit), non id temerario animi motu (quasi vento) incitatus facio, sed quod aut ipsi eas postulent, aut tanta mihi cum illis sit consuetudo ut apud eos infantiam meam aperire non pudeat. Ceterum quamvis horum neutrum acciderit, tamen crebra amicorum cohortatio hoc subinde suadentium sic me perpulit ut hasce qualescunque a me extorserit.

Quarum beneficio, auspicandae tecum amicitiae fores aperirem, nec tamen diffitebar me tam olim hoc studio et desiderio flagrasse, ex quo tu e Gallia in nostrum Lusitanum commigraveris; cumque omnes tui admiratione stupefacti in aedes tuas confluerent, audituri primam illam numquam satis laudatam orationem, quam ex facundiae tuae penu depromptam bona patriciorum adolescentiae parte considente felicissime habuisti. Nolui tantae rei spectaculo carere tantaque voluptate defraudari, sed continuo dimisso auditorum meorum conventu, illuc advolavi. Cumque prae multitudine ingedientium vix mihi pateret aditus, tandem ex angulo quodam tacitus auscultare coepi. Oratio itaque illa tanta animum dulcedine titillavit auresque demulsit ut unum ex illis oratoribus quos celebrat antiquitas audire mihi viderer. Quare inuriam meis studiis fieri putabam, si tecum, quem tantopere suspicio et admiror, nullam ineundae gratiae occasionem aucuparer.

Patere igitur, vir humanissime, desideriis meis compotem evadere; nam si me in amicitiae tuae particulam aliquam admiseris, polliceri possum me nec officio nec fide erga te, nec conservandae atque augendae amicitiae constantia, a quoquam tuorum superatum iri. Vale.

Jerónimo Cardoso sends greetings to António Pinheiro, Royal Councillor.

"Most eminent Sir, I know how irregularly I am acting, in prolonging my duty until today, except that it should have been discharged at this time, and in undertaking the duty of writing to you once again. However, if you examine the truth as an uncorrupted judge, you will decide that it was not a sign of carelessness nor of stupidity, but rather of prudence and of good judgement. For although I am clearly conscious of my weakness, as one blessed with nothing but mediocrity (yet tolerable, hopefully), I have always been afraid to call upon those men in particular, or provoke them with my writings, who are recommended both by their well-deserved fame and by their outstanding erudition, in case I should either burst like Aesop's frog, if keen to rival them or compare myself with them, or else reveal my poverty and rags, while thinking of showing off my resources or my riches. But if at some stage I have an opportunity to sent a letter to erudite men, as certainly does happen, I do not do so stirred up by a rash impulse of my mind (as if by a wind), but rather, because either they themselves ask for a letter, or I am such a close companion of theirs that I am not ashamed to reveal my lack of eloquence in their presence. However, although neither of the two has happened, yet frequent encouragement from my friends, who persuade me to do this straightaway, has put so much pressure on me that it has extorted this letter from me, such as it is.

Through the benefit of this letter, I may have a chance to start a friendship with you. I would not deny, however, that I have been burning with this intense desire for a very long time past, from the moment when you migrated from France to our Portugal, and when everyone poured into your home, stunned by their admiration for you, to listen to that first oration of yours, never praised highly enough, which you took from your store of eloquence, and delivered so successfully, with a good part of our young patricians sitting around you. I myself was unwilling to miss the spectacle of such a great event, or to be cheated of so great a pleasure, and so I at once dismissed my class of pupils and flew over to your place. Although an entry was barely open for me, thanks to the multitude of people going inside, yet from a corner I finally began to listen, in silence. That speech of yours aroused my mind with such pleasure and so charmed my ears that I thought that I was listening to one of those orators made famous by Antiquity. I thought that my studies would suffer harm if I did not look for an opportunity to begin a friendship with you, as I respected and admired you so highly.

And so, most learned of men, let yourself become a partner in my desires; for if you admit me into just a fraction of your friendship, I can promise you that not one of your friends will surpass me either in dutifulness or in fidelity towards you, or in the constancy of preserving and increasing our friendship. Farewell."

Pinheiro's Latin Poems honouring Francisco de Holanda

Included in Holanda's *Da Pintura Antigua*

Antonius Pinarius

Astra, quod in varias rerum distincta figuras
    quidve quod alliciat conditus orbis habet,
multiplici Pater omnipotens nisi singula cultu
    picta suum veluti proposuisset opus,
non secus ac tabulam si congessisset in unam   5
    omnia, sed propriis reddita quaeque locis.
Accidens pictura chaos distinxit in orbem.
    Orbi ea si desit, rursum erit ille chaos.

5 *congesisset* Vasc.   7 *distinxect* Vasc.

António Pinheiro

Because the foundation of the world has the stars separated
    into varied shapes of things, or what it might attract,
if the omnipotent Father had not put forward individual
    paintings with multiple decoration, as if his own work,
just as if he had collected everything together into one   5
    painting, with each thing assigned to its special place,
your Painting by appearing, separated chaos into a world.
    If the world were without it, that chaos will exist again.

Aliud Epigramma

Semina quis rerum picturae educta per artem
    ex indigesta mole rudique neget?
Ipse opifex rerum, verbum et sapientia patris,
    idem est aeterni lux et imago patris.

Satis

Another   Epigram

Who would deny that the seeds of things have been drawn
by the Picture's art from an undigested and raw mass?
The true creator of things, word and wisdom of the Father,
the same is the light and image of the eternal Father.

Enough

The second poem appeared in Sylvie Deswartes' interesting
and splendidly illustrated book *Ideias e Imagens em
Portugal na época dos Descobrimentos* (Lisboa, 1992).
Unfortunately there are several mistakes in her Latin
quotations, including the short epigram above, where she
reads *sapientiae* in the third line, which neither scans nor
makes any sense. Not that the Latin of Joaquim de
Vasconcellos (ed., Porto, 1918, p. 63) was much better, with
two errors in one poem, although he did print *sapientia*.
The two nouns *verbum* and *sapientia* are in apposition to
the artist (*opifex*). In Holanda's day, Latin was the common
language of the well-educated class. As a result any
scholars working on Renaissance writers today are sure to
need a reasonable knowledge of Latin grammar and
prosody, and in a few cases, as with Filelfo and Nunes, of
Ancient Greek also.

The vocabulary of the short poem is hardly 'neo-platonic',
as Deswartes recently argued (*"dois epigramas ..utiliza,
especialmente no segundo, um vocabulário de teor
neoplatónico e pictórico de tipo holandiano"*.) Rather, it is
a blend of Pinheiro's Classical learning and strict Catholic
orthodoxy. The second line is nicely modelled on Ovid's
Metamorphoses I.7 (where he describes the Creation):
[*chaos*] *rudis indigestaque moles*; Ovid also used *semina
rerum* for the elements, or atoms, imitating Lucretius.

If anything, the first poem is more Platonic, with chaos
being given forms through Holanda's *Pintura*. However,
unlike the nicely composed second poem, and his poem for
Nunes, the first one here is far too complex in its structure,

and not always good Latin. The second *quod* in line 2 is otiose, *habet* is awkward and *quid alliciat* ("what it may attract" or "draw to itself") is unclear. The two past unfulfilled conditionals with pluperfect subjunctives in lines 3-5 are far too heavy, and too prosaic for poetry, especially elegiac couplets, and they lack a proper apodosis (*distinxisset* would be expected, but would not scan). The poem is too convoluted to give all that much credit to its author; as a comparison with the mellifluous 12 verse poem dedicated to the same book by Pedro Sanches clearly shows (included by Joaquim Vasconcellos in his ed. of *Da Pintura Antigua*, pp. 62-3). Sanches was a talented Latin poet, whereas Pinheiro was a self-confessed admirer of Quintilian in his early years, and he seems to have had far more aptitude for Latin prose, especially its translation, sometimes at a very high speed, than for Latin poetry, unlike his fellow humanists at the Royal Court in Évora. Although he must have tutored Prince John while speaking Latin, Pinheiro chose Portuguese as his main language thereafter, both for public speaking and for his many prayers and sermons, some of them on momentous public occasions.

Appendix E

Coelho's Epigram on Nunes' *De Sphera*

The courtier and Christian poet, Jorge Coelho, dedicated this 10 verse epigram to Nunes' 1537 ed. of his *De Sphera*.

Georgii Coeli Epigramma

Qui cupis e terris arcana incognita caeli
    noscere et ignoto pandere vela mari,
en tibi, qui summum reserat sublimis Olympum.
    Per medios fluctus hoc duce tutus eris.
Haud mirum ingenii tot opes florere libello   5
    nobilis egregium condidit auctor opus.
Si clarum Alcidae durat per saecula nomen,
    quod caelum potuit sustinuisse humeris,
non minor et Petri dicenda est gloria Nonni,
    cuius mens terras, aequor et astra capit.   10

Epigram by Jorge Coelho

You who wish to learn the hidden secrets of Heaven from
    the earth, and to open your sails to an unknown sea,
look, an amazing man unseals highest Olympus for you.
    Led by him, you will be safe amid the waves.
It is no surprise that such a wealth of genius flowers in   5
    the book, a splendid work written by a noble author.
If the name of Hercules remains famous for centuries,
    for being able to hold up the sky on his shoulders,
the glory of Pedro Nunes should not be said to be less,
    whose mind holds the earth, the sea and the stars.   10

Coelho's elegant epigram well describes the talent and scientific originality of Nunes, whose discoveries were to prove so important for the extraordinary success of the

Portuguese navigators and crews (see vv. 2 and 4), as they charted dangerous straits and unknown shores. The Roman concept of *gloria* was very much alive in Coelho's day, and his relative Vasco da Gama had just been depicted as a second Aeneas in *The Lusiads* of Camoens. This glory was the special preserve of the *fidalgos* in warfare, making Coelho's claim for Nunes a very bold one, especially as he surpasses Hercules with his grasp of all aspects of Nature. The poem also hints at his ability as a teacher, and at the special quality of the *De Sphera*, a work by Nunes never translated into Latin. The poem is flattering, of course, and was written by a successful courtier, but it still suggests genuine admiration for the tutor of the royal Princes.

Jorge Coelho was also a close friend of Resende and of Jerónimo Cardoso.[142] His connection with Nunes is restricted to this epigram, but their paths must have crossed on many occasions. Coelho's success as a writer was in the field of Christian poetry, and he was renowned for his piety. For Nunes, a special interest in the Bible appears to have been a later development in his life, possibly triggered by the death of his friend and patron, Prince Louis. By contrast, Resende was a very reluctant courtier, and more a humanist than a devout priest, until the premature deaths of his two ex-pupils, Cardinal Alphonse and Prince Edward. During his later years, however, he became Chaplain and theological adviser to Cardinal Henry, the over-active Inquisitor General, and the one-time pupil and life-long friend and patron of Pedro Nunes. For all his religious bigotry, the Cardinal certainly attracted some brilliant intellects.

NOTES

1 Rebuilt at great expense by King John III, it still provides pure water today. His restoration of the Roman aqueduct was applauded in Latin poems written by several humanists attached to his Court, including one of 186 verses by Miguel da Silva, Cardinal of Viseu, for an English version of which see J. R. C. Martyn *André de Resende: On Court Life* (Peter Lang, Bern, 1990), pp. 205-210.

2 'Pedro Nunes: Classical Poet', *Euphrosyne* ix (1991), 231-270. For a far more accurate picture of the texts for these poems, especially the Greek ones, see J. R. C. Martyn & K. J. McKay, 'Pedro Nunes *Poemata: nonnulla corrigenda*' *Euphrosyne* xx (1992), 395-399. A preview of this book can be seen in my article 'The Teaching Manual of Pedro Nunes' in *Humanismo Português na Época dos Descobrimentos, Actas de Congr. Internacional* (Coimbra, 1993), pp. 275-280.

3 With my English version. When her fiancé repudiated his solemn vows, Guiomar's sense of honour and fierce indignation led to her slashing his face with a blade, before joining a nearby nunnery. It took place on 17 January 1578, only seven months before the death of her 76 year old father. The shock of this scandal probably accelerated the old man's demise. See Appendix B.

4 A full and authoritative biography is urgently needed, especially for scholars who cannot read Portuguese.

5 See Kenneth G. McIntyre, *The Secret Discovery of Australia* (Pan Books, 1988[2]), pp. 108-111, for Nunes' discoveries. His case for a Portuguese visit to southern Australia in 1522, terminated by a tempest at Warrnambool (to judge from the Dauphin Map), is a very convincing one, whether the 'Mahogany Ship' is found again or not. Hostile descriptions of this hulk by 'experts' have in fact proved to be reliable evidence for its Portuguese origin.

6 The *De Sphera* and *De arte atque ratione navigandi*.

[7] Mercator (or Gerard Kramer) must have been familiar with the works of Nunes, at least through their mutual friend, John Dee. In 1558 Dee wrote to Mercator saying that he had made Nunes the executor of his Will. See A. Fontoura da Costa, *A Ciênçia Náutica dos Portugueses na época dos descobrimentos* (Lisboa, 1958), p. 80. For the debt of Mercator to the double-cordiform projection of Oronce Finé, see Kenneth G. McIntyre *op. cit.* , p. 95.

[8] See C. R. Boxer *From Lisbon to Goa 1500-1750* (London, 1984), pp. 175-6. In Portugal, unlike England, the sailor rated well below the soldier in the social hierarchy, and most ship's officers were landsmen (Boxer *op. cit.* pp. 44-5).

[9] Even reaching the southern shores of Australia. See n. 5 above. In ch. 6 a fuller analysis of Nunes' discoveries and scientific inventions can be seen.

[10] Not even Camoens, the author of the epic on Vasco da Gama's discoveries, the *Lusiads*, the only Portuguese work of this period in the Penguin series (tr. W. C. Atkinson, 1952). For Camoens' debt to a Latin version of the Ines de Castro theme by André de Resende, see ch. 3 of my book *António Ferreira: The Tragedy of Ines de Castro* (Coimbra U.P., 1987; distrib. Classics Dept., Melbourne University).

[11] He is depicted on the 100 escudos coin holding the world in his hands. On the 200 escudos coin, Garcia de Orta holds a book in his right hand and a plant in the left. The two scientists were lecturers together at Lisbon University, although de Orta soon left for India, probably due to religious pressure. There his botanical and medical research resulted in his path-breaking book on *Colóquios dos simplices e drogas e cousas medicinais da India* (Goa, 1563). Ironically, both of these highly honoured scientific humanists were Jewish by birth, in a country where the converted Jews (*novos-christãos*) were the main victims of the Spanish Inquisition.

[12] Oronce Finé, professor of Mathematics at the Collège de France; see ch. 6.

[13] See ch. 6 for the impact of his *algébra* on contemporary and later mathematicians in Europe. The 187 problems in

his Spanish text would provide a good test for a modern class of first-year mathematics students.

14 In Portuguese, the 'nonio'.

15 Converted Jews for the most part. See notes 11 and 23. Cardinal Henry, Portugal's first active Inquisitor General, conducted its first *autos-da-fé* outside the 'Palace of the Inquisition' in Évora. Most of the Jewish bankers, doctors and businessmen escaped to the Low Countries, together with much of the commerce of Portugal (and of Spain).

16 Nunes used geometry and nautical theory, combined with practical navigation, to write a treatise on questions raised by Martim de Sousa, who had been in charge of the fleet that had headed for Brazil in 1531, to map its coasts with astronomical observations, returning in 1533.

17 The *Biblioteca Pública e Arquivo Distrital de Évora*. Luckily I had over two months to work on the miscellanies, barely enough time to scrutinize them all, especially with the library's closure for siestas, and unexpected holidays.

18 *Libro de Algebra en Arithmetica y Geometria*.

19 See Doreen Innes and Michael Winterbottom *Sopatros the Rhetor: Studies in the text of the Διαίρεσις Ζητημάτων* (B.I.C.S. Supplement, London, 1988). Sopater's table of στάσεις is on p. xi, and is compared with that of Hermogenes on p. 2. Sopater has an extra division, παραγραφή, and a different order (4, 6, 5 and 8, 10, 9). Hermogenes of Tarsus was greatly admired as a young sophist by Marcus Aurelius, and was a major figure in the Hellenistic development of rhetorical theory. His books on Στάσεις and Περὶ Ιδέων were basic texts until the Renaissance. Nunes is using the Aldine text of the scholia by Syrianus, Sopater and Marcellinus (*Rhetores Graeci*, 1509). Sopater was a late fourth century Athenian rhetor and Neo-Platonist who moved to Alexandria to teach and write. Besides his Διαίρεσις he published a prologomena on Aelius Aristides. For a preview of my work on this find, see my 'Lectures on Rhetoric given by Pedro Nunes at the University of Lisbon' *Euphrosyne* 23 (1995), 281-288.

[20] King John III gave orders for the University of Lisbon to be closed in 1536, and moved to Coimbra. Any opposition to this move was most unwelcome to the King, especially when it came as a petition signed by most of the Lisbon professors. A full list of senior staff can be seen in Jerónimo Cardoso's poem which praises the University's staff *Ad Olysipponensis Academiae Doctores*, included in his *Sylvae*, written in 1536, but not published until 1564 - well after the death of an unforgiving King. When Pedro joined the Lisbon staff, the new Rector, Jorge Fernandes, was an expert on Civil Law, as were Gonçalo Vaz Pinto and Francisco de Monçon. Pontifical Law was taught by Gabriel Dinis and Francisco Gentil, Medicine by Diogo Franco and Theology by Gonçalo de Santa Cruz. Jerónimo Clenardo, despite his legal background, seems to have looked after Classical literature. But for all intents and purposes the Lisbon University was a Law school, with minimal Medicine and Theology, and no professors of Grammar, Rhetoric, Logic, Mathematics, Philosophy or Science, until Pedro Nunes took over the chairs of Moral Philosophy, Logic and Metaphysics.

[21] A right hand, fingers only visible, beneath a rosette with five petals, similar only to parchment produced in Genoa from the early to mid-16th century. See C. M. Briquet *Les Filagranes* (Leipzig, 1923), under *Main*. The same watermark appears from folios 1^r to 18^v, but on folio 19^r a different watermark is visible, at the start of his 31 epitaphs for Prince Louis, 22 of which are in Latin and 9 in Greek. This must have been the cut-off point for his earlier compositions, written in Évora, most of them in Ancient Greek. See ch. 7.

[22] The family home of the very wealthy Portuguese ambassador to the Court of Charles V, Pedro Mascarenhas, the pupil and patron of Resende, was also at Alcácer do Sal, and there were probably links between Mascarenhas and Nunes (for example, through their common friend, Resende) but none are recorded. See note 51 below.

23 See notes 11 and 15 above. Not all of the New Christians who relied on Court patronage were safe. In his biography of Prince Edward, ch. 9, Resende exemplifies the Prince's rather malicious sense of humour with two stories about New Christians. A cheeky Paz, caught visiting a Rabbi's house, was treated by the Prince to a banquet of fatty bacon on the spit, bacon pie and bacon stew, until Paz vomited, and later the Prince pulled a glue-filled cap over his head. Failing to confess his sins, Paz was arrested and poisoned in the Inquisition's prison. A Jewish merchant then refused to supply clothes to the Prince, as it was the Sabbath. When they had been supplied, his payment was denied day after day when he called to collect it, as it was always some or other Christian Saint's day. The Prince's Jew-baiting is surprising, but no doubt reflected a popular bigotry. A Nunes (Pedro, *ut vid,*) is addressed by the Prince as 'one of his favourites', who had criticised him for excessive hunting and riding, in ch. 6 of the biography. For King Manuel's protection of his Jewish bankers, doctors and merchants, despite pressure from Spain, see my book *António Ferreira; The Tragedy of Ines de Castro* (Coimbra, 1987), pp. 114-118.

24 The sons of his daughter Isabel da Cunha (died 1621) and João Pereira de Sampaio. Matias Pereira was imprisoned in Coimbra (1623-1631) and Pedro Nunes Pereira in Lisbon (1623-1632), yet they both survived their ordeal. There was no friendly Prince left to protect them.

25 See Joaquim de Carvalho 'Pedro Nunes - Mestre do Cardeal Infante D. Henrique' *A Cidade de Évora* VII, 21-2, pp.11-12, where he produces documents listing payments to Nunes as 'fysico' ('doctor') by both Princes. In making his interesting discoveries in Évora's library (Cod. cvii/1-29), he came remarkably close to discovering the *algébra*.

26 Or *Liber de nuptiis Mercurii et Philologiae*, a prose and verse *satura* written between 410 and 429 AD in Carthage. The *Trivium*, consisted of Grammar, Dialectic and Rhetoric.

27 See Carvalho *op. cit.* for the Portuguese account of this onerous programme.

28 Resende was the only one of these three friends and fellow tutors to publish his Latin poems, mainly the longer ones; they were part of his *Opera* ed. Arnoldus Mylius, 2 Vols., Cologne 1600, and amount to over 5,000 verses. I recently discovered an equally large body of his Latin poems, but mainly short ones (over 5,000 vv) which are awaiting publication. For a recent biography of this brilliant scholar, theologian and poet, see my book *André de Resende: On Court Life* , ch. 1. Clenardo, renowned as an extempore Latin poet, was fluent also in Hebrew, Greek and Arabic. See my 'Resende, Clenardo and Erasmus' *Euphrosyne* xxi, 1993, 375-388 and Alphonse Roersch *Correspondance de Nicolas Clénard* (2 Vols., Brussels, 1940/41).

29 For Filelfo's success with Greek lyric poetry, see Diana Robin *Filelfo in Milan* (Princeton U.P., 1991), pp. 8-9 & 122-137. Filelfo had studied Ancient Greek under Cardinal Bessarion in Constantinople. See ch. 7 for Nunes' Greek and Latin poetry, and similarities between him and Filelfo.

30 For his generous stipends, see Joaquim de Carvalho's article in *A Çidade de Évora*. He was survived by his wife and four daughters and possibly two sons, none of them distinguished (except the infamous Guiomar).

31 See note 21 above. The new watermark has five stylized fingers with the letters WI above, inside an archway with a crown on its summit.

32 The *Oratio de Sapientia*. See Domingues, Gabriel de Paiva *Oração de André de Resende pronunciada no Colégio das Artes em 1551* (Coimbra, 1982).

33 For his public oration and complaints of onerous teaching duties there, see my book *On Court Life* pp. 40-43.

34 Jerusalem, 1971, Vol. 12, Column 1273.

35 As in the Nunes biography in the *Grande Enciclopédia Portuguesa e Brasileira* (Lisboa, 1925-1960), under Nunes (Pedro) p. 60, col. 1: '*O original foi primitivamente escrito em português, circa 1534-1535, vertendo-o mais tarde ... para castelhano.*' Joaquim de Carvalho 'Pedro Nunes - Mestre do Cardeal Infante D. Henrique' *A Cidade de Évora*

VII, 21-22, 10, argues that the Infante *viu* ('saw') the original manuscript while still his student. In his notes on the *Libro do Algebra* in Nunes *opera* vol. vi (Lisboa, 1940), p. 420 he assigned its *redacção primitiva* to 1535-1537; but admitted that *tudo o mais é incerto*..('all the rest is uncertain'). See also R. Hooykaas *Voyages of Discovery in 16th century Portuguese Science and Letters* (Amsterdam, 1979), p.141:'About the same time (1535-36) Nunes wrote in Portuguese a book on algebra which, however, was published only 30 years later.' Not a 'book', but the right dates. Rodolfo Guimarães in his *Sur la vie et l'oeuvre de Pedro Nunes* (Coimbra, 1915) also suggested "il a été écrit en 1532 ou 1535 ..cette réduction est perdue et seule la dédicace au Cardinal Don Henrique, datée de Lisbonne le 1er déc. 1564, a été conservée dans la langue original".

36 For his discoveries & scholars' dated theories, see ch. 6.

37 See Guimarães *op. cit.* p. 71, quoting Wallis' view (Oxford, 1693, p. 67).

38 Their value for modern students of Algebra is stressed by Henri Bosmans S. J. *L'Algèbre de Pedro Nuñez* (Coimbra, 1908) ch. 1.

39 See Joachim de Carvalho, *op. cit.*, 427-8, and the *Grande Enciclopédia Portuguesa e Brasileira* , under Nunes, p. 53.

40 *Nesta opulentissima cidade de Lixboa, onde tanto negotio ha desde extremo oriente e occidente, e ilhas do mar Oceano, e onde el Rey nosso senhor tem quarenta contadores de sua fazenda.*

41 *Por esta causa, vendo enquanto seia util para ho uso dos homens esta arte que trada dos numeros e medidas.*

42 As argued by Henri Bosmans *op. cit.*, 435-6. Copies of it are in few libraries today, outside Portugal. To justify his choice of Castilian, Nunes adapted a text translated by him in his early days at Lisbon University, Aristotle's *Politics* vii.i.1323 b 10, "With any goods of the soul, the more abundant it is, the more useful it must be": 'the more common and universal a good thing is, the more excellent it is'. See R. Hooykaas *op. cit.* p. 50.

43 His pedagogic *Tratado em defensam da carta de marear* (1537) became *De arte atque ratione navigandi* soon afterwards. His other major scientific works in Latin were the *De Crepusculis* (1542) and *De erratis Orontii Finaei* (1546).

44 See R. Hooykaas, *op. cit.* pp. 50-51.

45 See Guimarães, *op. cit.* p. 66, note 2. Some ocopies in city libraries may have been lost during the two World Wars.

46 George Buchanan (1506-1582) made use of Latin for his beautifully written but quickly out-dated astronomical work *De Sphaera*.

47 See Maria de Brito, *op. cit.* pp. 15-17.

48 It would also have given him a chance to see his elder brother Rodrigues once again. He had been serving in India since André's childhood, and was owed a large sum of back pay, as André pointed out, enough to help him to support the two orphans in his charge. Whether Castro ever made good the money owed to Rodrigues, and whether he passed it on to André, is unknown. He makes it clear in his letter, however, that it was his sickness, and his heavy family responsibilities (with two or three sisters to maintain, as well as the two orphaned children), that were key factors stopping him from joining João de Castro, as I have shown in my book *André de Resende: On Court Life.* pp. 29-32, where an English translation of his letter is included. He had earlier published his *Epitome rerum gestarum in India* in 1531. In her book *Ideias e Imagens em Portugal na época dos Descobrimentos* (Lisboa, 1992), Sylvie Deswartes has a very interesting section (see p. 42 especially) on the Indian antiquities, and Resende's refusal to accept the challenge. She does not seem to recognize his family ties and illness at Viana, when Castro might have used his personal charm to persuade him. To Deswartes, Holanda was better able to record India in full. She still credits Janus Vitalis with the famous poem *De Roma*, although it seems certain that it was originally written by André de Resende (see my article 'André de Resende - Original author of *Roma Prisca*' *Bibl. d'Human. et Ren.* li,

1989, 407-411). With a Martial epigram 'found in Rome', it seems fairly certain that Resende visited the eternal city.

49 A list of Easters for 1511-1557 was provided for me by Prof. Aires Nascimento; April 13th appears beside 1533. On the Perpetual Calendar, 1533 begins on the 25 March

50 For the imaginary story of their heroic choice of death before dishonour in 238 AD, see the *Legenda Aurea* by Jacobus de Voragine.

51 See my 'André de Resende and the 11,000 Holy Virgins' *Humanitas* xxxix-xl (1987-8), 197-209. The three prayers for Responsa appeared at the end of Resende's *Translatio Responsae* (Venice, 1532). Mascarenhas paid heavily for these relics by financing the rebuilding and redecoration of their Abbey. Resende's prayers are devotional masterpieces. Mascarenhas' father was in charge of the Spanish horses of Kings John II and Manuel, but died in 1501. For Pedro's illustrious career, culminating with his death as Viceroy of India, see note 3 in the article above, and ch. 2 in my *André de Resende : On Court Life* that describes a sumptuous banquet staged by him in Brussels, with Charles V as guest-of-honour. Among other things, he was an expert in fortress design, and acted as the Emperor's adviser. Most experts in designing and repairing fortresses at this time were Italians, as at Mazagão. See my recent book *The Siege of Mazagão* (Peter Lang, New York, 1994).

52 See Brito, Maria Fernanda de, *Pedro Nunes na Tipografia de Quinhentos* (Coimbra,1979), p. 5, quoting Gomes Teixeira.

53 The Spanish version of Nunes' *algebra* is yet to be translated into English or any other European language, despite the value of its problems for students today.

54 *op. cit.* pp. 420-426.

55 *História das Matemáticas em Port.* (Lisboa, 1934), p. 154.

56 The French original appears in n. 1 on page 5 of Bosman's *L'Algèbre de Pedro Nunes* : "I'y encores ouî dire de Pierre None, Mathematicien de Lisbonne en Portugal, qu'il l'auoit aussi traictée (l'algèbre) en son langage Espagnol; mais ie n'ay veu son liure."

[57] The *'son' langage* above certainly suggests that he supposed Pedro Nunes to be Spanish by birth.

[58] See *Pedro Nunes Obras Completas* (Acad. das Cíençias, Lisboa, 1946) Vol vi, *Notas* p. 413. For Vinet's comment on Nunes' *De erratis Orontii Finaei*, see *ibid.*, p. 414.

[59] Johannes de Sacrobosco (John Hollywood) studied at Oxford before becoming Prof. of Maths in Paris, dying there in c. 1244. His treatise was based on Ptolemy's *Almagest*, and ran to 40 editions between 1472 and 1647. He was one of the first to use Arabic writings on astronomy.

[60] "Qu'un mien ami Pero Nunez, Cosmographe du Roi de Portugal Iehan le Tiers, publia et fit imprimer a Coimbre Université de Portugal. l'an M.D.XLVI." See Nunes *Obras* Vol.I, pp. 293-4.

[61] Translated into Latin six years later as *Iacobi Peletarii Cenomani, De Occulta parte numerorum quam Algebram vocant, Libri duo* (Paris, 1560).

[62] Peletier also published an *Ars Poetica* in 1555, of interest to the poetic Pedro Nunes.

[63] See David E. Smith, *History of Mathematics* 2 Vols, (New York, 1958), vol. I.

[64] See Guimarães, *op. cit.*, p. 73.

[65] See Bosmans, *op. cit.*, pp. 5-6. Nunes' comments are on folio 224[r].

[66] He may have kept the notations used in his first *álgebra*, even after Stifel's work had appeared.

[67] See Bosmans, *op. cit.*, pp. 13-14. He mentions a work by a Sicilian abbot, Francisco Maurolyco, *Arithmeticorum libri duo nunc primum in lucem editi* (Venice, 1575), who seemed to support Némore's approach.

[68] Many of his rules and formulae could be valuable for modern classes in algebra. See Lemos, Victor Hugo Duarte de, *Notas e comentários* on Pedro Nunes *Obras*, Vol. VI, Lisbon, 1946, p. 498: 'A fórmula então usada para resolver algèbricamente as equações consideradas é, porém, a que ainda hoje se usa'. See also notes 13 and 38 above.

[69] *Op. cit.*, p. 17.

[70] This suggests that Nunes had refuted Finé's theories before leaving for Coimbra, again very early in his mathematical career, possibly in 1533-35.

[71] *De erratis Orontii Finaei, Regii Collegii Mathematicarum Lutetiae Professoris, qui putavit inter duas datas lineas, binas medias proportionales sub continua proportione invenisse, circulum quadrasse, cubum duplicasse, multangulum quodcumque rectilineum in circulo describendi artem tradidisse, et longitudinis locorum differentias aliter quam per eclipses lunares, etiam dato quovis tempore manifestas fecisse, Petri Nonii Salaciensis Liber unus.* It was re-edited in Coimbra in 1571, and in Basle in 1592.

[72] See note 7 for Mercator's debt to Finé's double-cordiform projection. Like Nunes, Finé saw the importance of mathematics for practical problems of navigation. Mercator virtually copied his projection.

[73] See above for Vinet's use of this work by Nunes, whose translation was a main source of astronomical information for Camoens, especially for Canto X.

[74] Georg von Purbach (1423-1461), professor at Vienna, is generally regarded as the first modern astronomer.

[75] *Ut occasionem aliquam nancisceret excusandi me quod interpretationem Vitruvii tam diu sim moratus: nam prae adversa valetudine inchoatum opus et supra quam dimidiatum non absolui.* (p. 2)

[76] On page 78, section 38.

[77] See Joaquim de Carvalho, 'Uma obra inédita de Pedro Nunes: Defensão do Tratado da rumação do globo para a arte de navegar' *Revista Filosófica* Coimbra, 1 (1951), 176-180.

[78] Published in his *Tratado da Sphera* (Lisboa, 1537), his only scientific work entirely in Portuguese. For his use of geometry to solve errors in the variations of the corrections with the altitude, see Luís de Albuquerque, 'The Historical Background to the Cartography and the Navigational Techniques of the Age of Discovery ..', *Estudos de História*, v (1977), 12-13. See pp. 19-21 for João de Castro's

contribution to the solving of cartographical problems, as in his 1538 voyage to the Indies.

[79] See my 'Pedro Nunes - Classical Poet', *Euphrosyne* xix (1991), 231-270. In *Euphrosyne* xx, 1992, 395-9, the texts of the poems, especially the ones in Greek, were corrected extensively by me and my colleague, Dr Ken McKay.

[80] See the 1991 article above pp. 233-4, 236-237 and 250-251. His relationship with Jerónimo Cardoso can be seen in Appendix D.

[81] Published in Lisbon in 1542, after also being sent to Cardoso for his opinion. See Appendix D. Pinheiro's poem was omitted in the next edition (Coimbra, 1573).

[82] See Luís de Matos, *Les Portugais à l'Université de Paris* (Coimbra, 1950), pp. 61-4, for his enrolment, and pp. 147-50 for his dedicatory letter to his teacher, Diogo de Gouveia. In it he says "me puerum complexus es", before entrusting him to Iacobum Lodoicum (Jacques Louis Strébée), author of *De Electione et Oratoria Collocatione Verborum* (Paris, 1538). For a very brief biography of Pinheiro, see Américo da Costa Ramalho, 'Pinheiro, António', *Verbo-Enciclopédia Luso-Brasileira de Cultura*, Lisboa, 15, 1973, 114-115. He gives no date of birth.

[83] Reprinted three times, in Paris and Venice.

[84] Luís de Matos first thought so (*Les Portugais* p. 72) but added a correction on pp. 174-5, to show that Frei André was in Naples in November 1541, coming from Flanders. However, he may have visited Évora before his trip to Italy. It seems that his memory was very prone to lapses.

[85] The date in Ramalho's biography; he noted the links with Sanches, Vaz and Pires.

[86] See Elisabeth Feist Hirsch, *Damião de Gois: The Life and Thought of a Portuguese Humanist, 1502-1574* (The Hague, 1967), p. 164. See pp. 181-2, for her very brief biography of Pinheiro.

[87] For his growing concern over Jesuit power, see Hirsch, *op. cit.* pp. 209-210. In 1565 he visited and reformed the University of Coimbra.

88 The text reads *pungit* ('pierces', or 'torments') but this reading makes little sense, and a subjunctive is needed. The imagery of *pingat* ('paints') suits the context, and is a likely printing error.

89 Similar in theme to Juvenal *Satire* VII, which starts with an appeal to the Emperor as the only hope for the Muses.

90 In his well documented *Les Portugais à l'Université de Paris*.

91 See my book *António Ferreira: The Tragedy of Ines de Castro*, pp. 15-35.

92 I hope to publish a work on the totally unexpected death of the College's Principal in the next year or two. The evidence on his last hours can be seen in the College pharmacist's report to the Inquisition, and it has been closely examined by two French doctors, who saw arsenic as the only possible explanation.

93 See my book *On Court Life*, pp. 33-37. Another slave was freed by him in his Will.

94 Fetched by Resende from a Chair he had just taken up at the University of Salamanca, Clenardo was offered 100,000 *reís*, and given 50 gold coins on his arrival in Évora. He shared accommodation with Resende, whom he had taught Hebrew while Resende was studying in Paris, 1526-8.

95 Although he pleaded poor, Resende had been well rewarded by the King and his successors, as his Will reveals. See my book *On Court Life*, pp. 33-36.

96 Nunes' generous salaries and endowments are well brought out by Joaquim de Carvalho in his article in *A Cidade de Évora* VII, 21-22. See also A. Fontoura da Costa *op. cit.* pp. 24-5.

97 In Nunes' very interesting dedication of his *D e Crepusculis* to John III, he explains how it was initiated by an inquiry from Prince Henry. "Ten years ago (1531) you took care that he (Henry) should be taught mathematics by me, and he learnt most diligently and in a short time the arithmetical and geometric elements of Euclid, the mechanics of Aristotle, all cosmography and the use of instruments, some early ones and some invented by me for

the art of navigation.   But if he had studied them longer, he certainly would have become perfect in mathematics; but he had to be ordained and concentrate on the splendid study of theology.   Yet each day he asks for a really knotty, difficult, problem, demanding geometrical demonstrations from me, having no time for them himself.   A few days ago he asked about the longitude of shadows in different climates." Dedicating his *algebra* to Henry was no surprise.
98 In the next year or so I hope to publish Resende's *Vida do Infante Dom Duarte* with an English translation, as *The Life and Death of Prince Edward*. His account of Prince Edward's brilliant mind and very pious life and premature death is very like Nunes' elegiacs in honour of Prince Louis.   Each of these humanists and fellow tutors lost a dear friend and powerful protector, and both ended up having to rely on the Princes' brother, Cardinal Henry, and young Sebastian, for their future patronage and protection.
99 António Franco, *Synopsis* p. 45, 27. xi. 1555.
100 As in a poem on *The Eucharist* : 'If the desire for the fairest land of all for your ancestry / gnaws at you, a land Christ richly adorned for the holy, / a land that Jesus Christ illuminated with a spring of beauty ...'.
101 See Odette Sauvage *L'Itinéraire Érasmien d'André de Resende (1500-1573)* pp. 102-137, for her notes and French translation of this very Erasmian oration.   A Portuguese version by Miguel Pinto de Meneses was included in the edition of this important oration by Artur Moreira de Sá, published in Lisbon, 1956.
102 This suggests the complex problems of algebra, the key to his many practical scientific discoveries.
103 For his 10 divisions of rhetorical lectures, see   ch. 3.
104 Aristotle had 10 'categories', or 'predicables', reduced to 8 (*Phys*: 5, 1, 13), thereby classifying all of the ways in which assertions could be made on a subject (in Grammar, the 8 'parts of speech', Subst., Adj., Advb. and Verb, active, passive, trans., intrans. and conditional).   The modest *Isagoge* by Porphyry the Phoenician became a best-seller during the controversy over universals, especially when

adapted by Abelard. From Aristotle's four primary predicables (property, genus, accident and definition), Porphyry removed the last, and added 2 of his own, species and difference. See Edward W. Warren, *Porphyry the Phoenician: Isagoge* (Toronto, 1975).

105 *De Verborum Conjugatione Commentarius* (Lisboa, 1540), f. 30.

106 Taken from my book *On Court Life,* pp. 192-3.

107 For Filelfo's Greek poetry, see Diana Robin, *op. cit.,* pp. 122-137, and Enrico V. Maltese, 'Osservazioni critiche sul testo dell' epistolario greco di Francesco Filelfo' in *R e s Publica Litterarum. Studies in the Classical Tradition* 11 (1988), pp. 207-213. As Robin points out, his poems were "among the first Neo-Greek poems to be written by an Italian humanist".

108 See *Resende: Vida do Infante Dom Duarte,* ch. 19.

109 His marriage in 1523, when he was only 21 years old, was to Dona Guiomar de Arias, daughter of Pedro Fernandez Arias, a Spanish Christian living near Salamanca.

110 According to tradition, Nicodemus was the very wealthy young man who asked Jesus what he should do to ensure a place in Heaven, whereas it was Joseph of Aramathea who gave his tomb as Jesus' resting-place after his crucifixion.

111 [os] for grammatical correctness, as in *hecháisos.*

112 The last 2 words are underlined, and uncertain; the sentence is incomplete.

113 literally 'puff of breath'.

114 This implies that a first point has been treated, on the feeding of the 5,000.

115 An error for *tanta,* it seems.

116 The prep. is superfluous, as *pequeño* is an adjective.

117 A Portuguese word for the Castilian *espuertas.*

118 From Isaiah 40. 6.

119 The first Council of Nicaea met in 325, to produce a Creed to combat various anti-Trinitarian doctrines.

120 The "world' is used collectively for its inhabitants.

[121] The religious notes were for the benefit of Nunes only, and he knew the sequel to this lead-in.

[122] The Feast of the Annunciation (Lady Day) was set for March 25, but it moves to the second Tuesday after Easter when it falls within Lent.

[123] *çufre* in the manuscript.

[124] From Psalm 45. 13 (in the Vulgate, 44. 14, with *gloria* for *decus*).

[125] In the original *dios* was read, but crossed out.

[126] Matthew, ch. 20, *ut vid.*, ending "the last shall be first, the first last".

[127] The guardian of Esther, bride of King Ahasuerus. His favourite minister, Haman, intended to hang Mordecai on gallows he had erected for him. The Jew's past services and Esther's influence led to Haman and his sons being hung instead, and Mordecai being highly honoured.

[128] From Luke 1. 28, describing the Annunciation.

[129] *i. e.* after the Annunciation.

[130] Luke 1. 42-45.

[131] Corrected from *trabajando* in the manuscript.

[132] The first word of a new paragraph, it seems, on the next folio, now lost.

[133] See A. Fontoura da Costa, *op. cit.,* pp. 14-15.

[134] See note 24 above.

[135] For *figens in virgine vultus* and *lilia albe rosa* see *Aeneid* XII, 68-70.

[136] In lines 57-64, over half the words are taken from Virgil's famous storm simile in *Aeneid* I. 148-153. The apparently young poet goes overboard in his imitation.

[137] His *Oratio Pro Rostris* was published in 1550 in Coimbra, and was edited with a translation by Dr Justino Mendes de Almeida (Lisbon, 1965).

[138] I hope to publish them with an English translation in 1996 or 1997, with the help of the Portuguese Government.

[139] *Hieronymi Cardosi Lusitani Epistolarum Familiarum Libellus* (Olysipone, 1556). Only one of the letters gives the year (to John III, dated 1555), but Nunes' first Latin

work, *De Crepusculis Liber Unus*, dedicated to King John III on 18 October 1541, seems to have been sent to Cardoso soon afterwards, to be published in Lisbon in 1542.

140 Astrologer in Later Latin (Mathematician in Cicero).

141 Pinheiro had dedicated a Latin epigram to applaud the same Latin text as that sent to Cardoso; see ch. 7, *ad init.* He also sent two epigrams to Francisco de Holanda for his *D a Pintura Antigua.* These can be seen after the letter, translated into English.

142 Several of Cardoso's letters were written to Coelho, and Resende teased his very devout and all too proud friend about his name ('rabbit' in Portuguese) and his ancestors (who included Vasco da Gama). See my book *On Court Life* pp. 22-4. The epigram can be seen on page 241 of Vol. I of the Academy's edition of Nunes' works, with some misleading punctuation (full-stops at the ends of lines 2, 7 and 8) plus *author* in 6 and *Alcide* in 7.

Bibliography

Albuquerque, Luís de   'The historical background to the Cartography and the navigational techniques of the age of discovery, with special reference to the Portuguese' *Estudos de Historia* , Coimbra, v, 1977, 1-24

------- 'Science et Humanisme dans la Renaissance Portugaise' in *Actes du xxi colloque Int. d'Études Humanistes. Tours, Juillet 1978*, Paris, 1984, pp. 419-435

------- *A projeção da náutica portuguesa quinhentista no Europa* (Coimbra, 1972)

------- 'Pedro Nunes e Diogo da Sá' *Memórias da Academia das Ciências de Lisboa* xxi, 1976-7, 339-357

Amorim, Diogo Pacheco de *Pedro Nunes : Subsidios para a sua biografia* (Coimbra, 1935)

André, Carlos Ascenso *Diogo Pires: Antologia poética* (Coimbra, 1983)

Aquarone, Jean-Baptiste *D. João de Castro, gouverneur et vice-roi des Indes Orientales (1500-1548)* (Paris, 1968, 2 vols)

Azcona, J. M. López de *Quatro centário da publicação da Tratado de Sphera de Pedro Nunes* (Lisboa, 1938)

Bensaude, Joaquim *L'astronomie nautique au Portugal à l'époque des grandes découvertes* (Bern, 1912)

Bosmans, Henri *L'Algèbre de Pedro Nuñez* (Coimbra, 1908)

------- 'L'Algèbre de Jacques Peletier du Mans' *Rev. des Questions Scientifiques*, Brussells, 61, 1907, 117-73

------- 'Le "De Arte Magna" de Guillaume Gosselin' *Bibliotheca Mathematica*, Leipzig, 7, 1906-7, 44-66

------- 'Sur le 'Libro de Algebra' de Pedro Nuñez' *Bibliotheca Mathematica* 8, 1907-8, 154-169

Boxer, C. R. *From Lisbon to Goa 1500-1750* (London, 1984)

------- *Fidalgos in the Far East 1550-1770* (London, 1968)

Brandão, Mário *O Colégio das Artes* (Coimbra, 1924-33)

------- *A Inquisição e os Professores do Colégio das Artes* (Coimbra, 1948)

------- *O processo na Inquisição de Mestre João da Costa* (Coimbra, 1944)

Bricker, Charles and Tooley, R. V. *A History of Cartography* (London, 1969)

Brito, Maria Fernanda de Pedro *Nunes na Tipografia de Quinhentos* (Coimbra, 1979)

Briquet, Charles M. *Les Filagranes* (Leipzig, 1923)

Camoês, Luís de *Os Lusíadas*

Carvalho, J. M. Teixeira de *Homem de Outros Tempos* (Coimbra, 1924)

Carvalho, Joachim de *Notas* on Nunes' *álgebra* in *Obras* Volume VI, 422-437

------- 'Pedro Nunes - Mestre do Cardeal Infante D. Henriques' *A Cidade de Évora* vii, 1950, 4-13

Cassen, Lionel *The Ancient Mariners* (New York, 1959)

Castro, João de *Obras Completas* ed. A. Cortesão and L. de Albuquerque (Coimbra, 1971) Vol. I *Roteiro de Lisboa a Goa em 1538* Vol. II *Roteiro do Mar Roxo em 1541*

Cipolla, Carlo M. *Guns, Sails and the Early Phases of European Expansion 1400-1700* (New York, 1965)

Collingridge, George *The First Discovery of Australia and New Guinea* (Pan Books, Sydney, 1982[2])

Cortesão, Armando *Nautical Science and the Renaissance* (Coimbra, 1974)

------- *History of Portuguese Cartography* (Coimbra, 1971)

Costa, A. Fontoura da *Pedro Nunes (1502-1578)* (Lisboa, 1969[2])

------- *A Ciencia nautica .. .. descobrimentos* (Lisboa, 1958)

------- *A Marinharia dos descobrimentos* (Lisboa, 1960)

------- *D. João de Costa: Roteiros (I - II)* (Lisboa, 1940)

Costa, José Fernandes 'O cosmógrafo Pedro Nunes, D. Guiomar sua filha e inexactidão da qualidade de poeta, que ao primeiro foi recentemente atribuída' *Bol. de Seg. Classe da Acad. das Ciências de Lisboa* 54, 1923, 134-142

Deswartes, Sylvie *Ideias e Imagens em Portugal na época dos Descobrimentos* (Lisboa, 1992)

Dictionary of Scientific Biography ed. Gillespie, Charles C. (New York, 1974) Vol. X

Dimmock, L. 'The Lateen rig' *The Mariner's Mirror* 32, 1946, 35-41

Domingues, Gabriel de Paiva *Oração de André de Resende pronunciada no Colégio das Artes em 1551* (Coimbra,1982)

Elliot, J. H. *Imperial Spain: 1469-1716* (Pelican, 1965)

Escobar, T. Martin *Sobre el "Libro de Algebra en Arithmetica y Geometria" de Pedro Nunes* (Madrid, 1932)

Farinha, Bento José de Sousa *Colleçam das obras portuguesas do António Pinheyro* 2 vols. (Lisboa, 1784/85)

Gaullieur, E. *Histoire du Collège de Guyenne d'après un grand nombre de documents inédits* (Paris, 1874)

Godinho, Vitorino Magalhães 'Le Tournant Mondial de 1517-1524 et l'Empire Portugais' *Studia* 1, 1958, 184-199

Goertz, R. O. W. 'The Portuguese in Cochin in the mid-sixteenth century' *Studia* 49, 1989, 5-38

Grande Enciclopédia Portuguesa e Brasileira 40 vols, (Lisboa, 1925-1960) For Nunes (Pedro), see pp. 53-65

Guerreiro, José Manuel 'André de Resende e o Humanismo em Portugal' *A Cidade de Évora* 37-38, 1955-56, 5-53

Guimarães, Rodolfo *Les mathématiques en Portugal* (Coimbra, 1909[2])

------- *Sur la vie et l'oeuvre de Pedro Nunes* (Coimbra, 1915)

Hirsch, Elisabeth Feist *Damião de Gois: The Life and Thought of a Portuguese Humanist, 1502-1574* (The Hague, 1967)

Historia Mathematica for 'Nuñez Salaciense' *v.* pp. 160-162

Hooykaas, R. *Humanism and the Voyages of Discovery in 16th cent. Portuguese Science & Letters* (Amsterdam, 1979)

Humanismo Português na Época dos Descobrimentos (Actas, Congresso International Coimbra, Out.1991, Universidade de Coimbra, 1993)

Innes, Doreen and Winterbottom, Michael *Sopatros the Rhetor: Studies in the text of the* Διαίρεσις Ζητημάτων (B.I.C.S. Supplement, London, 1988)

Jugé, Clément *Jacques Peletier du Mans. Essai sur sa vie, son oeuvre, son influence* (Paris, 1907)

Kayserling, Meyer *História dos Judeus em Portugal* (São Paulo, 1971)

Kennedy, G. A. *Greek rhetoric under the Christian Emperors* (Princeton, 1983)

------- *The Art of Persuasion in Greece* (Princeton, 1963)

Klein, Jacob *Greek Mathematical Thought and the Origin of Algebra* tr. Eva Brann (M.I.T. Press, 1968)

Kline, Morris *Mathematical Thought from Ancient to Modern Times* (O.U.P., 1972)

Kuhn, Thomas S. *The Structure of Scientific Revolutions* (Chicago, 1962)

Lane, Frederick C. 'The Economic Meaning of the Invention of the Compass' *Amer. Hist. Rev.* lxviii, 3,1963, 605-617

Lemos, Victor H. D. de *Notas e Comentários* on Pedro Nunes' *álgebra* in *Obras* Vol. VI (Lisboa, 1946), pp.471-498

McKintyre, Kenneth Gordon *The Secret Discovery of Australia* (Pan Books, Sydney, 1982²)

Maltese, Enrico V. 'Osservazioni critiche sul testo dell' epistolario greco di Francesco Filelfo' in *Res Publica Litterarum. Studies in the Classical Tradition* 11 (1988), pp. 207-213

Manguin, Pierre-Yves 'A mid-17th collection of *roteiros* for Asian waters' *Studia* 48, 1989, 187-211

Marcus, G. J. 'The Mariner's Compass, its Influence upon Navigation in the Later Middle Ages', *History* xli, 1, 1956

Martins, Augusto 'Pedro Nunes e a algébra' *A Aguia* Porto, 2, 1, 1912, 23-6

Martins, Isaltina das Dores Figueiredo *Bibliografia do Humanismo em Portugal no Século XVI* (Coimbra, 1986)

Martyn, John R. C. *António Ferreira: The Tragedy of Ines de Castro* (Coimbra, 1987, distrib. by the author)

------- *André de Resende: On Court Life* (Lang, Bern, 1990)

------- *The Siege of Mazagão: A Perilous moment in the conflict between Christianity and Islam* (Lang, N.Y. 1994)

------- 'Clenardo, Resende and Erasmus' *Euphrosyne* xxi, 1993, 375-388

------- 'Pedro Nunes - Classical Poet' *Euphrosyne* xix, 1991, 231-270

------- 'The Teaching Manual of Pedro Nunes', in *Humanismo Português na Época dos Descobrimentos* (Coimbra, 1993), pp. 275-280

------- 'André de Resende - Original Author of *Roma Prisca* ' *Bibliothèque d'Human. et Renaissance* 51, 1989, 407-411

Martyn, J. R C. and McKay, Kenneth J. 'Pedro Nunes : *Poemata*: nonnulla corrigenda' *Euphrosyne*, xx, 1992, 395-9

Marques, J. M. da Silva *Descobrimentos Portugeses* 2 vols (Lisboa, 1944-9)

Matos, Luís de *Les Portugais à l'Université de Paris entre 1500 et 1550* (Coimbra University, 1950) Lettre-dédicace d'António Pinheiro à Gouveia l'Ancien pp. 147-150

------- *L'Expansion Portugaise dans la Litterature Latine de la Renaissance* (Lisboa, 1991)

Mocatta, Frederic D. *The Jews of Spain and Portugal and the Inquisition* (New York, 1973)

Mota, A. Teixeira da 'Méthodes de Navigation et Cartographie Nautique dans l'Océan Indien avant le XVIe siècle' *Studia* 11, 1963, 49-91

Nemésio, Vitorino *Vida e obra do Infante D. Henrique* (Porto, 1959)

Nougarède, M. P. 'Qualités nautiques des navires arabes' *Studia* 11, 1963, 95-122

Nunes, Pedro *Obras Completas* (Acad. das Cíençias, Lisboa, 1946) esp.*Tratado que o doutor Pero Nunez fez sobre certas duvidas da navegação* and *Tratado que o doutor Pero Nunez fez em defensam da carta de marear* (Lisboa, 1537)

Ortis, Antonio Dominguez *The Golden Age of Spain, 1516-1659* tr. James Casey (London, 1971)

Pastor, J. Rey *Los Matemáticos españoles del siglo xvi* (Madrid,1926)

Phillips-Birt, D. *A history of seamanship* (New York, 1971)

Prestgate, Edgar *The Portuguese Pioneers* (London, 1966^2)

Ramalho, Américo da Costa *Estudos sobre a Época do Renascimento* (Coimbra, 1969)

------- 'Pinheiro, António', *Verbo-Enciclopédia Luso-Brasileira de Cultura* , Lisboa, 15, 1973, 114-115

------- *Latim Renascentista em Portugal* (Coimbra, 1985)

------- *Para a História do Humanismo em Portugal* (I.N.I.C., Coimbra, 1988)

Révah, I. S. 'Les origines de Jerónimo Cardoso, auteur du premier dictionaire portugais imprimé' *Boletin da Acad. das Ciéncias de Lisboa,* Lisboa, 36, 1964, 277-279

Robin, Diana *Filelfo in Milan* (Princeton U.P., 1991)

Roersch, Alphonse *Correspondance de Nicolas Clénard* (3 vols., Brussels, 1940/1941)

Rose, Paul Lawrence *The Italian Renaissance of Mathematics* (Genève, 1975)

Roth, Cecil *A History of the Marranos* (New York, 1959)

Russell, G. A. *Greek Declamation* (Cambridge, 1983)

Sá, A. Moreira de *O infante Dom Henrique e a Universidade* (Lisboa, 1960)

------- 'Jerónimo Cardoso' *Humanitas* 30, 1983, 252-75

Sanceau, Elaine *Cartas de D. João de Castro* (Lisboa, 1954)

------- *D. João de Castro* (Paris, 1956²)

Sarana, António José *Historia da Cultura em Portugal* (Lisboa, 1965)

Sauvage, Odette *L'Itinéraire Érasmien d'André de Resende (1500-1573)* (Paris, 1971)

Schurhammer, Georg 'S. Francisco Xavier e a sua época' *Studia* 12, 1963, 7-28

Silva, Luciano Pereira de *As obras de Pedro Nunes, sua cronologia bibliográfica* (Coimbra, 1925)

Smith, David Eugene *History of Mathematics* 2 Vols (Dover, New York, 1958)

Sölver, C. V. and Marcus, G. J. 'Dead Reckoning & the Ocean Voyages of the Past' *The Mariner's Mirror* xliv, 1958, 18-34

Sousa, Morais e *A sciência náutica dos pilotos portugueses nos séculos XV e XVI* (Lisboa, 1924)

Souza, T. O. Marcondes de 'A astronomia náutica na época dos descobrimentos marítimos' *Revista de História*, São Paulo, 20, 1960, 41-63

Struilc, D. J. *A Source Book in Mathematics,1200-1800* (Harvard U.P.,1969)

Taylor, A. H. 'Carrack into Galleon' *The Mariner's Mirror* 36, 2, 1950

Taylor, E. G. R. *The haven-Finding Art: a history of navigation from Odysseus to Captain Cook* (London, 1956)

Teixeira, F. Gomes *História das Mathemáticas em Portugal* (Lisboa, 1934)

Teyssier, Paul 'Jerónimo Cardoso et les origines de la lexicigraphie portugaise' *Bull. des Études Port. et Brésil.*, Lisboa, 41, 1980, 7-32

Themudo, Marilia A. de L. Monteiro 'Pedro Nunes - eminente matématico português' *O Primeiro do Janeiro*, Oporto, 15. viii. 78, 3

------- 'Pedro Nunes - Matématico insigne: geómetra, algebrista, cosmógrafo e pioneiro da inventiva portuguesa' *O Primeiro do Janeiro* 22. xi. 78, 12

Turner, Gerard L'E. 'An Astrolabe attributed to Gerard Mercator, c. 1570' *Annals of Science* , 50, 1993, 402-443

Vasconcellos, Joaquim de (ed.) *Francisco de Holanda: Da Pintura Antigua* (Porto, 1918)

Vinet, Élie *Joannes de Sacrobosco, Sphaera, emendata* (Paris, 1561)

Walter, Jaime 'O Infante D. Henriques e a medicina' *Studia*, 13-14, 1964, 31-39

Warren, Edward W. *Porphyry the Phoenician: Isagoge*: (Toronto, 1975)

Waters, David W. *The Art of Navigation in England in Elizabethan and Early Stuart Times* (London, 1958)

------- *The Rutters of the Sea* (Yale U.P., 1967)

Zeuthen, H. G. *Histoire des Mathématiques dans l'Antiquité et le Moyen Âge* (Paris, 1902)

INDEX

Names of the authors of the modern works listed in the Bibliography above are not included in this Index.